EVOLUTION AND THE FALL

Evolution and the Fall

Edited by

William T. Cavanaugh *&* James K. A. Smith

WILLIAM B. EERDMANS PUBLISHING COMPANY
GRAND RAPIDS, MICHIGAN

Wm. B. Eerdmans Publishing Co.
2140 Oak Industrial Drive N.E., Grand Rapids, Michigan 49505
www.eerdmans.com

Published 2017
Printed in the United States of America

23 22 21 20 19 18 17   1 2 3 4 5 6 7

ISBN 978-0-8028-7379-8

Library of Congress Cataloging-in-Publication Data

A catalog record for this book is available from the Library of Congress

Contents

IV. Reimagining the Conversation: Faithful Ways Forward

Foreword

One thing is clear: as a culture we no longer have the place or patience needed to tackle difficult issues and questions. Where does one go to consider and discuss the pressing issues of the day, to explore questions whose answers are not readily or easily found?

At The Colossian Forum, we believe that difficult questions—like the ones discussed in this book—belong to (and so belong in) the community that is the church. We confess that we have been given, in the church and its teachings, everything we need to engage divisive cultural questions in ways that both extend the tradition faithfully and deepen human flourishing. What's more, in the church's practices of worship we've been given a common space in which we might take up these difficult issues. We need not rush through to an easy answer, but, rooted in liturgical time, we can, together with all the saints, address questions as occasions to manifest love of God and neighbor.

The Colossian Forum exists to equip leaders to transform cultural conflicts into opportunities for spiritual growth and witness. This book is one example of the kind of dialogue we have in mind. Over the course of three years, ten scholars from a variety of disciplines gathered with The Colossian Forum to consider and address one particularly difficult question: If humanity emerged from nonhuman primates (as genetic, biological, and archaeological evidence seems to suggest), then what are the implications for Christian theology's traditional account of origins, including both the origin of humanity and the origin of sin?

As we considered this question together, we did so not from within the practices of the secular academy but from within the shared confessions and practices of the Christian faith. We took into account the sense of the faithful and the pressures on local pastors as well as the most recent scholarship. We considered the capacity to glorify Christ and receive his Word as a central evaluative criterion of truth, thus equipping us to engage the developments in genetics and paleoanthropology with fresh eyes and renewed hope.

More than a collection of essays tilling the same old tired ground, this book represents an innovative conversation with some of today's brightest Christian minds, seeking to build up the church in both truth and love. It's just this sort of conversation that you're invited to join.

Thanks to the editors, Bill Cavanaugh and Jamie Smith, for skillfully guiding such a diverse group of scholars on such an unusual adventure, drawing together diverse trains of thought into a compelling vision for the future of this conversation. Their wisdom, ecclesial hearts, and, perhaps most importantly, sense of humor and delight in the experience made it an honor to accompany them and the team all along the way.

MICHAEL GULKER
President, The Colossian Forum
September 2016

Acknowledgments

While we hope this book contributes to the future of engagements between science and theology, for those of us involved it will also be a reminder of a season spent together as friends, collaborators, and co-pilgrims. This was all made possible by the gracious hospitality of the Colossian Forum, whose vision and mission informs this project, and whose virtues and practices were experienced by all of us involved in the project. We are deeply grateful to Michael Gulker, President; Rob Barrett, Director of Forums & Scholarship; and the inexplicably kind and omnicompetent Andy Saur for their support, encouragement, patience, and inspiration.

We are also grateful to the busy scholars who agreed to be part of this project. We asked them for a lot. Most significantly, we asked them for a not insignificant commitment to time together, despite many other demands on their time. We hope their experience of this collaboration was as life-giving for them as it was for us.

This project was underwritten by the generous support of the BioLogos Foundation's Evolution and Christian Faith grant program (which, in turn, was funded by the John Templeton Foundation). We would like to express our appreciation to Deborah Haarsma and Kathryn Applegate for their support, wisdom, and flexibility.

Contributors

William T. Cavanaugh is professor of Catholic studies and director of the Center for World Catholicism and Intercultural Theology at DePaul University. His degrees are from Notre Dame, Cambridge, and Duke. He has published numerous articles and seven books, most recently *Field Hospital: The Church's Engagement with a Wounded World* (Eerdmans, 2016). Other books include *Torture and Eucharist: Theology, Politics, and the Body of Christ* (Blackwell, 1999) and *The Myth of Religious Violence: Secular Ideology and the Roots of Modern Conflict* (Oxford, 2009). His books have been published in French, Spanish, Polish, Norwegian, Arabic, and Swedish.

Celia Deane-Drummond is currently full professor in theology at the University of Notre Dame and director of the Center for Theology, Science, and Human Flourishing. Her research interests are in the engagement of theology and natural science, including specifically ecology, anthropology, genetics, and evolution. She has published or edited twenty-five books and over two hundred scientific and scholarly articles or book chapters. Her most recent books include *Future Perfect*, ed. with Peter Scott (Continuum, 2006, 2nd ed. 2010); *Ecotheology* (DLT/Novalis/St. Mary's, 2008); *Christ and Evolution* (Fortress/SCM, 2009); *Creaturely Theology*, ed. with David Clough (SCM, 2009); *Religion and Ecology in the Public Sphere*, ed. with Heinrich Bedford-Strohm (Continuum, 2011); *Animals as Religious Subjects*, ed. with Rebecca Artinian Kaiser and David Clough (T&T Clark/Bloomsbury, 2013); *The Wisdom of the Liminal: Human Nature, Evolution and Other Animals* (Eerdmans, 2014); and

Technofutures, Nature, and the Sacred: Transdisciplinary Perspectives, ed. with Sigurd Bergmann and Bronislaw Szerszynski (Ashgate, 2015).

Darrel R. Falk is professor of biology (emeritus) at Point Loma Nazarene University in San Diego, CA, where he has been since 1988. He is the former president of the BioLogos Foundation from 2009 to 2012 and currently serves as BioLogos's Senior Advisor for Dialog. He earned a doctorate in genetics from the University of Alberta and did postdoctoral work at the University of British Columbia and the University of California, Irvine before beginning his career on faculty at Syracuse University. He has given numerous talks about the relationship between science and faith, and is the author of *Coming to Peace with Science: Bridging the Worlds Between Faith and Biology* (IVP, 2004).

Joel B. Green is provost, dean of the School of Theology, and professor of New Testament interpretation at Fuller Theological Seminary. He has written or edited more than forty-five books, including *Conversion in Luke-Acts: Divine Action, Human Cognition, and the People of God* (2015); *Why Salvation?* (2013); *Ears That Hear: Explorations in Theological Interpretation of the Bible* (co-editor, 2013); *Practicing Theological Interpretation* (2011); and *Body, Soul, and Human Life: The Nature of Humanity in the Bible* (2008). He is the editor of the New International Commentary on the New Testament, and editor-in-chief of the *Journal of Theological Interpretation*. He also serves on the editorial boards of the journals *Theology and Science* and *Science and Christian Belief*. Green has been elected to membership in both the Studiorum Novi Testamenti Societas (SNTS) and the International Society for Science and Religion (ISSR).

Michael Gulker is president of The Colossian Forum in Grand Rapids, MI. Michael has a longstanding interest in the intersection of faith and culture and how both thrive best when rooted in worship. A native of West Michigan, he studied philosophy and theology at Calvin College, has a divinity degree from Duke Divinity School, and is an ordained Mennonite pastor. Before coming to The Colossian Forum, Michael served as pastor of Christ Community Church in Des Moines, IA. He and his wife Jodie have two young children.

Peter Harrison is an Australian Laureate Fellow and Director of the Institute for Advanced Studies in the Humanities (IASH) at the University of Queensland. Prior to this he was the Idreos Professor of Science and Religion and Director of the Ian Ramsey Centre at the University of Oxford. He has published extensively in the area of intellectual history focusing on the early modern period and on the historical relations between science and religion. He is the author of several books, including *The Bible, Protestantism, and the Rise of Natural Science* (Cambridge, 1998); *The Fall of Man and the Foundations of Science* (Cambridge, 2007); and, most recently, *The Territories of Science and Religion* (Chicago, 2015), a revised version of his 2011 Gifford Lectures.

J. Richard Middleton is professor of biblical worldview and exegesis at Northeastern Seminary, in Rochester, NY. He is currently a BioLogos Theological Fellow and serves as adjunct professor of Old Testament at the Caribbean Graduate School of Theology (Kingston, Jamaica). He was president of the Canadian Evangelical Theological Association, 2011–14. Middleton's books include *A New Heaven and a New Earth: Reclaiming Biblical Eschatology* (Baker Academic, 2014); *A Kairos Moment for Caribbean Theology*, ed. with Garnett Roper (Pickwick, 2013); *The Liberating Image: The Imago Dei in Genesis 1* (Brazos, 2005); and two coauthored volumes with Brian Walsh: *Truth Is Stranger Than It Used to Be: Biblical Faith in a Postmodern Age* (IVP, 1995); and *The Transforming Vision: Shaping a Christian World View* (IVP, 1984). He is currently working on a book for Baker Academic called *The Silence of Abraham, The Passion of Job: Explorations in the Theology of Lament*. His books have been published in Korean, French, Indonesian, Spanish, and Portuguese.

Aaron Riches is a joint faculty member of the Instituto de Filosofía Edith Stein and the Instituto de Teología *Lumen Gentium* in Granada, Spain, where he teaches theology at the Seminario Mayor San Cecilio. He received his PhD in Systematic Theology at the University of Nottingham. He is the author of *Ecce Homo: On the Divine Unity of Christ* (Eerdmans, 2016) and has published articles in a number of academic journals, including *Modern Theology, Telos, Communio, The International Journal of Systematic Theology*, and *Nova et Vetera*.

James K. A. Smith is professor of philosophy at Calvin College, where he holds the Gary and Henrietta Byker Chair in Applied Reformed Theology and Worldview. The award-winning author of *Who's Afraid of Postmodernism?* and *Desiring the Kingdom*, his most recent books include *Imagining the Kingdom* (2013); *Discipleship in the Present Tense* (2013); *Who's Afraid of Relativism?* (2014); *How (Not) to Be Secular: Reading Charles Taylor* (2014); and *You Are What You Love: The Spiritual Power of Habit* (2016). His popular writing has appeared in magazines such as *Christianity Today, Books and Culture*, and *First Things* and periodicals such as the *New York Times, Wall Street Journal*, and

USA Today. Smith is also a Senior Fellow of Cardus and serves as editor of *Comment* magazine.

Brent Waters is Jerre and Mary Joy Stead Professor of Christian Social Ethics and director of the Jerre L. and Mary Joy Stead Center for Ethics and Values at Garrett-Evangelical Theological Seminary in Illinois. Brent received his Doctor of Philosophy at the University of Oxford and both his Doctor of Ministry and Master of Divinity at the School of Theology at Claremont. Some of his most recent publications include *Christian Moral Theology in the Emerging Technoculture: From Posthuman Back to Human*; *This Mortal Flesh: Incarnation and Bioethics*; *The Family in Christian Social and Political Thought*; *From Human to Posthuman: Christian Theology and Technology in a Postmodern World*; and *Just Capitalism: A Christian Ethic of Globalization*.

Norman Wirzba is professor of theology and ecology at Duke Divinity School and research professor at Duke's Nicholas School of the Environment. He has published *The Paradise of God: Renewing Religion in an Ecological Age*; *Living the Sabbath: Discovering the Rhythms of Rest and Delight*; *Food and Faith: A Theology of Eating*; and *Making Peace with the Land* (coauthored with Fred Bahnson). His most recent books are *From Nature to Creation: A Christian Vision for Understanding and Loving Our World* and *Way of Love: Recovering the Heart of Christianity*. He also has edited *The Essential Agrarian Reader: The Future of Culture, Community, and the Land* and *The Art of the Commonplace: The Agrarian Essays of Wendell Berry*. Professor Wirzba serves as general editor for the book series Culture of the Land: A Series in the New Agrarianism, published by the University Press of Kentucky, and is on the Executive Committee of the Society for Continental Philosophy and Theology.

Beyond Galileo to Chalcedon

Resources for Reimagining Evolution,
Human Origins, and the Fall

WILLIAM T. CAVANAUGH AND JAMES K. A. SMITH

This book addresses a set of problems that arise from the encounter of traditional biblical views of human origins with contemporary scientific theories about the origin of the human species. The scientific theories are, of course, a moving target; new evidence is unearthed, and different theories are frequently proposed, attacked, defended, and discarded. Nevertheless, there is a broad scientific consensus on some key issues that fits uneasily with the biblical tradition and cannot be ignored by theologians and the wider church. The scientific consensus points to the evolution of humans from primates. It indicates that humans emerged in a group, not an original pair. And the emergence of humans from primates seemingly leaves little room for an original historical state of innocence from which humanity suffered a "Fall." What then to do with biblical accounts of human origins and the doctrinal reflections of the Christian tradition on the Fall and original sin? Must we either relegate the biblical accounts to the category of "myth," or ignore the science of evolution? The chapters of this volume address the questions of human origins in detail. In this introduction, we address the wider and prior question of how Christians should approach the intersection of Christian doctrinal traditions with knowledge from outside those traditions.

From Galileo to Chalcedon

Christian convictions can generate "good" problems. For example, it is precisely biblical convictions about the goodness of creation and human culture-making that propelled scientific exploration of God's world, rightly seeing science as yet another vocation that can honor the Creator. But as we pursue that vocation, rooted in these biblical convictions, we encounter challenges: sometimes it seems the "book of nature" is telling us something different from what we've read in the book of scripture. And so we find ourselves in what philosopher Charles Taylor describes as a "cross-pressured" situation: our dual commitment to the authority of scripture and the affirmation of science places us in a space where we seem to encounter two different, competing accounts of human origins.

Some describe this as another "Galilean" moment, a critical time in history where new findings in the natural sciences threaten to topple fundamental Christian beliefs, just as Galileo's proposed heliocentrism rocked the ecclesiastical establishment of his day. This parallel is usually invoked in the face of genetic, evolutionary, and archeological evidence about human origins and development that presses against traditional Christian understandings of human origins. Since we now tend to look at the church's response to Galileo as misguided, reactionary, and backward, this "Galilean" framing of the new origins debate does two things: First, it casts scientists—and those Christian scholars who champion such science—as heroes and martyrs willing to embrace progress and enlightenment. Second, and as a result, this framing of the debate associates concern with Christian orthodoxy as backward, timid, and fundamentalist.

This Galileo analogy is a loaded one; it assumes a paradigm in which science is taken to be a neutral "describer" of "the way things are" whereas theology is a kind of bias, a "soft" *take* on the world that has to face up to the cold, hard realities disclosed to us by the natural sciences and historical research. Christian scholars and theologians who (perhaps unwittingly) buy into this paradigm are often characterized by a deference to "what science says" and become increasingly embarrassed by both the theological tradition and the community of believers who are not quite as eager to embrace scientific "progress." The result is that the Christian theological tradition is seen to be a burden rather than a gift that enables the Christian community to think through such challenges and questions well.

This book questions this construal of our situation. We agree that the church is at a critical juncture in the history of Christian thought—that issues around the historicity of Adam (and related issues of original sin) are crucial, difficult questions that the church must face. But we believe that before we can "solve" the tensions at the intersection of scripture and science when it comes to human origins, we need first to "hit the pause button," as it were, and consider just *how* the Christian community can work through such issues. And we believe that the "Galilean" metaphor is unhelpful and unproductive in this regard precisely because it already biases the conversation in an unhelpful way. Instead of fostering theological imagination, this approach tends to assume that the issues are settled and we just need to "get with the program"—which usually requires relinquishing some key theological convictions.

In contrast to this "Galilean" framing of the issues, we believe that Christian scholars can find an older model and paradigm in the ancient resources of Chalcedon. As Mark Noll has argued in *Jesus Christ and the Life of the Mind*, Christian scholarship is not rooted in merely "theistic" claims—and should certainly not be rooted in a functional deism. Rather, the proper place for Christians to begin serious intellectual labor "is the same place where we begin all other serious human enterprises. That place is the heart of our religion, which is the revelation of God in Christ."[1] Noll's point isn't just pious invocation of Jesus; rather, as he goes on to show, what's of interest in Chalcedon is the way the church navigated contemporary challenges with a theological imagination that was able to retain core Christological convictions while at the same time taking seriously the "science" (natural philosophy) of the day. A "Galilean" approach might have simply said: "Look, based on our current philosophical knowledge, it's impossible to affirm that someone is both human and divine. So you have to resolve this tension in one direction or the other: *either* Jesus is human *or* he is divine. He can't be both." But of course that is just the approach that Chalcedon refused. Instead, feeling the tension and challenge, the Council of Chalcedon exhibited remarkable theological imagination and generated what is now one part of the heritage of the church: the doctrine of the hypostatic union—that in the one person of Christ subsist two natures, divine and human. This is not a theological development that could have been anticipated before the church worked through the issues.

What if we thought of ours not as a "Galilean" moment, but a "Chalcedonian" opportunity? For some—often those who pose the question in "Galilean" terms—the choice seems clear: if humanity emerged as a result of human evolution, then there couldn't have been one Adam. And perhaps even more importantly: if humanity emerged from primates, then it seems that there could never have been a "good" creation or "original righteousness"—which would also mean that there was no "Fall" from a prior inno-

---

1. Mark Noll, *Jesus Christ and the Life of the Mind* (Grand Rapids: Eerdmans, 2011), xii.

cence. If we are going to affirm an evolutionary account of human origins, it would seem we need to give up on the doctrine of sin's origin and original sin.

But are things so clear? Have we yet created the space to exercise our theological imaginations on these issues as they did at Chalcedon? Have we properly appreciated what's at stake in these questions—how the threads of orthodox Christian theology are woven together, and how pulling on a loose thread might mar the entire tapestry? Could there be ways to think our way *through* this "cross-pressured" situation that, like Chalcedon, affirms the parameters of orthodoxy while taking seriously contemporary challenges? The goal is not to solve or escape the tensions and cross-pressures via a strategy that simply eliminates one of the elements of the challenge (whether the relevant science or traditional Christian doctrine). Rather, we embrace the cross-pressure as an impetus for genuine, yet faithful, theological development.

Imagination Takes (Liturgical) Practice

Creative and constructive theological work requires faithful imagination. But that requires two things: time and worship. We need *time* to train and stretch our imaginative muscles; time to ruminate on issues and opportunities; time to listen and contemplate; and above all, time to pray. So the cultivation of *faithful* imagination also requires bathing and baptizing the imagination in the cadences of the biblical story—which is precisely the goal of Christian worship. Thus the cultivation of constructive theological imagination begins with liturgical formation.

Behind this book is not only a set of theological convictions; the matrix of this project was also a community of prayer, worship, and friendship that provided both time and discipline for us to imagine otherwise. This book is the scholarly fruit of a three-year experiment—sponsored by the Colossian Forum on Faith, Science, and Culture—that gathered a multidisciplinary and ecumenical team of leading scholars to pursue a communal research program on evolution, the Fall, and original sin, addressing one of the most pressing questions of the day: if humanity emerged from nonhuman primates—as genetic, biological, and archeological evidence seems to suggest— then what are the implications for Christian theology's traditional account of origins, including both the origin of humanity and the origin of sin? The integrity of the church's witness requires that we constructively address this difficult question.

But our methodology is as central to our project as the topic. Following the suggestion of Mark Noll in *Jesus Christ and the Life of the Mind*, we embrace the church's ancient wisdom in the Council of Chalcedon as a model and template for how faithfully to grapple with contemporary challenges. In the Christological orthodoxy that emerged from the early church—the Apostles' Creed, Nicea, Chalcedon—we see the body of Christ taking seriously the challenges of the day (from "natural philosophy," i.e., science) while at the same time retaining (indeed formulating) the parameters of Christology. The result is a feat of Spirit-led imagination, giving us the formulations that we confess to this day.

We also believe the resources for such theological imagination are carried in the liturgical heritage of the church—in the worship practices and spiritual disciplines that enact the biblical story in ways that seep into our imagination. We think it is crucial that scholars working on the forefront of these issues be immersed in these practices as the "imagination station" for creative theorizing. Thus our research program gathered our team for times of reflective retreat. These practices were not just pious adornments of our intellectual labor but reservoirs for faithful Christian scholarship. The church needs a similar feat of Spirit-led imagination today to face questions of human origins and the origin of sin.

Christology in Practice

Echoing the animating convictions of the Colossian Forum, this research project is rooted in two related convictions:

First, we believe that the Christian intellectual tradition is uniquely "carried" in the practices of Christian worship. Thus we see liturgy, worship, and common practices of prayer as a central, formative resource for thinking well as Christians. It is in the prayers and worship of the church that we are immersed in the Word and our imaginations are located in God's story. If we need theological imagination to grapple with difficult issues, then the practices of Christian worship (and related spiritual disciplines) are fuel for such imagining and creative theological work. Thus we consider intentional liturgical formation to be the *sine qua non* for rigorous Christian scholarship.[2]

---

2. These claims and convictions are unpacked in much more detail in James K. A. Smith, *Desiring the Kingdom: Worship, Worldview, and Cultural Formation* (Grand Rapids: Baker Academic, 2009) and *Imagining the Kingdom: How Worship Works* (Grand Rapids:

Second, we believe that the Christian theological heritage, rooted in the Word and articulated in the creeds and confessions of the church, is a gift, not a liability. This doesn't mean that the orthodox theological heritage is simply a deposit to be repeated and repristinated. Rather, our theological heritage provides an invaluable foundation for building new theological models that address our increased knowledge about the natural world. Indeed, while some frame our situation as a distinctly "modern" one—a replay of that Galilean moment—we are convinced that the church has "been here before," well before Galileo. Indeed, we believe there is much wisdom to be found by going beyond Galileo to Chalcedon and other resources in the church fathers and medieval doctors of the church. Our sensibility, one might say, is an "ancient-future" one: we believe the church will find gifts to help it think through contemporary challenges by retrieving the wisdom of ancient Christians. The goal is not to simply repeat ancient formulations while sticking our heads in the sand with respect to these challenges; rather, we believe that the contemporary church—and contemporary Christian scholars—can learn much from the "habits of mind" that characterized ancient scholars like Athanasius and Augustine.[3]

Thus our research project is oriented not only by Christological convictions but by what we might describe as Christological *practice*, centering our intellectual work in the spiritual disciplines and worship practices of the church as incubators for our theological imagination. It is in these practices that we absorb a core conviction in our bones: that in Christ all things hold together (Col. 1:17). If virtues are the result of disciplines, intellectual virtue is also the fruit of discipline and practice. Thus our research team gathered not only for intellectual exchange but for common intellectual *formation*. This is also why *time* was so central to our project. We gathered over a three-year period, and spent a week together when we gathered, because we needed time together to pursue formation in common, and we needed time for our theological imaginations to percolate in the face of such difficult questions at the intersection of human origins and Christian faith.

The result, we hope, is a book that reflects the uniquely *communal and ecclesial* aspects of our project. The chapters that follow are not simply the results of discrete research agendas; they reflect a deep commonality and

---

Baker Academic, 2013). See also Nicholas Wolterstorff, "Christology, Christian Learning, and Christian Formation," *Books and Culture* 18, no. 5 (September/October 2012): 22–23.

3. For suggestions along these lines, see Peter J. Leithart, *Athanasius* (Grand Rapids: Baker Academic, 2011), and Timothy George, ed., *Evangelicals and Nicene Faith: Reclaiming Apostolic Witness* (Grand Rapids: Baker Academic, 2011).

unity that bubbled up from communal prayer and a common "canon" of ancient wisdom. The common voice of this book reflects the common prayer that nourished it. We also hope the fruit of our research exhibits a maturity that reflects the shared gifts of communal, embodied collaboration, forging a common sense of conviction, bringing the gifts of diverse disciplinary expertise to the table, and thus charitably pressing one another from those perspectives. We see such practiced friendship as the way to truth.[4]

Always Reforming: Theological Development in a Confessional Tradition

Assertions of confessional orthodoxy and recognition of theological development are not mutually exclusive. So we shouldn't be too hasty to assume that any and all assertions of confessional orthodoxy necessarily stem from backward, defensive stances that would merely repristinate historic creeds and confessions. Or, to put this otherwise, claims that certain proposals or conclusions fall outside the parameters of confessional orthodoxy do *not* assume that there cannot be legitimate theological development within a confessional tradition. Debates about human origins and the Fall need something of a "meta" account of theological development within a confessional tradition that honors the *dynamic* nature of such development without abandoning the *boundary-marking* function of a confessional tradition. To sketch this, we will extrapolate from Alasdair MacIntyre's account of how a tradition's practices are revised and extended.

Let's take the Christian tradition to be a "tradition" in MacIntyre's sense: "A tradition is an argument extended through time in which certain fundamental agreements are defined and redefined."[5] By the "Christian tradition" we refer to a catholic theological heritage, catalyzed by the scriptures, carried in the practices of Christian worship, and articulated in the ecumenical creeds (Apostles' Creed, Nicene Creed, etc.). This is already a living tradition

4. David Burrell, *Friendship and Ways to Truth* (Notre Dame: University of Notre Dame Press, 2000); see also his encapsulation of this point in "Friendship in Virtue Ethics," http://www.colossianforum.org/2012/07/12/glossary-friendship-in-virtue-ethics/.

5. Alasdair MacIntyre, *Whose Justice? Which Rationality?* (Notre Dame: University of Notre Dame Press, 1988), 12. In what follows, we are going to treat "practice" and "tradition" as roughly synonymous. That is not precisely how it works in MacIntyre—a tradition is carried in a community of practice; however, there is a symbiotic relation between the two that permits treating them as roughly synonymous.

with expanding layers of articulation, expansion, and revision internal to the tradition.

Now, as a "tradition" in the MacIntyrean sense, the Christian tradition is "carried" in a community of practice, which, roughly speaking, is "the church" (not the academy). And what characterizes such a practice, according to MacIntyre, is precisely the creative extension of the tradition by the community of practice. In other words, a tradition requires a dynamic of creative repetition rather than mere repristination. MacIntyre puts it this way:

> We are apt to be misled here by the ideological uses to which the concept of a tradition has been put by conservative political theorists. Characteristically such theorists have followed Burke in contrasting tradition with reason and the stability of tradition with conflict. Both contrasts obfuscate. For all reasoning takes place within the context of some traditional mode of thought, transcending through criticism and invention the limitations of what had hitherto been reasoned in that tradition; this is as true of modern physics as of medieval logic. Moreover when a tradition is in good order it is always partially constituted by an argument about the goods the pursuit of which gives to that tradition its particular point and purpose.
>
> So when an institution—a university, say, or a farm, or a hospital—is the bearer of a tradition of practice or practices, its common life will be partly, but in a centrally important way, constituted by a continuous argument as to what a university is and ought to be or what good farming is or what good medicine is. Traditions, when vital, embody continuities of conflict. Indeed when a tradition becomes Burkean, it is always dying or dead.[6]

A tradition will sometimes have to confront its own limitations, such as its own failure to articulate a coherent response to new scientific evidence and theories. The tradition "lives on" just insofar as the community of practice reappropriates the tradition *creatively* but also *faithfully*, and those two dynamics are not mutually exclusive. So a tradition is not merely a restatement of past formulations; *as* a tradition—especially as a tradition that enlivens a community of practice—new "performances" aim to "extend" the tradition. These (re)performances develop, refine, improve, and extend the tradition.

6. Alasdair MacIntyre, *After Virtue*, 2nd ed. (Notre Dame: University of Notre Dame Press, 1984), 221.

And part of that extension will include internal critique. In other words, it is of the very essence of a tradition to debate what constitutes "the tradition"—and especially what constitutes a "faithful" extension of the tradition. So part of the tradition is debating and revising the goals of the tradition.

However, like improvisation in jazz, such debate and internal critique are normed by the tradition.[7] There is a dynamic of authority that is also at work in this process of extension. Indeed, MacIntyre emphasizes that "[t]o enter into a practice is to enter into a relationship not only with its contemporary practitioners, but also with those who have preceded us in the practice, particularly those whose achievements extended the reach of the practice to its present point. It is thus the achievement, and *a fortiori* the authority, of a tradition which I then confront and from which I have to learn."[8] So being part of a tradition, being involved in the dynamics of extension and reform, comes with a price of admission, *viz.*, submission to the authority of the tradition.

Traditions are not closed off from the rest of the world. Traditions are constantly in conversation with other traditions, and what a tradition learns from outsiders is vital to making a tradition live. Christians, for example, should be grateful for the gifts that sciences like biology have given us. Although a tradition must always be ready humbly to learn from those outside the tradition, it is the tradition that yields its own internal criteria for what *counts* as a "faithful" extension of the tradition. In other words, what "counts" as a reason or warrant or evidence or a "good move" in this game is tethered to the heritage of the tradition.[9] This doesn't mean there is no room for innovation or creative extension, but it does mean that in order for a "move" to *count* as an extension it will have to be judged as faithful to the tradition.[10] And this is an inherently social, communal project of discernment: it is the community of practitioners—the community of those who have submitted

7. See Samuel Wells, *Improvisation: The Drama of Christian Ethics* (Grand Rapids: Brazos, 2004).

8. Wells, *Improvisation*, 194. See also MacIntyre, "Epistemological Crises, Dramatic Narrative, and the Philosophy of Science," in *Paradigms and Revolutions: Applications and Appraisals of Thomas Kuhn's Philosophy of Science*, ed. Gary Gutting (Notre Dame: University of Notre Dame Press, 1980), 54-74.

9. Cp. Robert Brandom's account of the "pragmatics" of reasoning and his almost ethnographic account of rational discourse as the "giving and taking of reasons." For a discussion, see James K. A. Smith, *Who's Afraid of Relativism? Community, Contingency, Creaturehood* (Grand Rapids: Baker Academic, 2014), 115-49.

10. So, for example, new moves in the tradition are not primarily based on their "relevance" or their "compatibility" with other regnant paradigms.

to the tradition—who judge whether a "new" move is a creative extension of the tradition, or whether such a move has broken the rules and is really playing a new game.[11] "The spirits of the prophets are subject to the prophets" (1 Cor. 14:32).

This seems an apt description of how a confessional tradition faithfully extends the tradition, "reformed, but always reforming."[12] In an important sense, discernment about "faithful extensions" is (scandalously) entrusted to the people of God, the priesthood of all believers—which, in many Christian communities, is guided by episcopal authority but in all includes the laity. There is a deep affirmation of, and trust in, the "sense of the faithful" (Newman) and the operation of the Spirit in leading the community into truth.[13] This is *not* meant to be a recipe for repristination or stubborn repetition; it is rather a dynamic for discerning what counts as a "faithful extension" of the tradition.

It is important to recognize this dynamic because it highlights something of a "clash of epistemologies" that characterizes the current debate, or at least a tension between two very different *epistemes*. Our guilds tend to have an "encyclopedic"[14] approach that sees knowledge's advance as a straight line of progress and development, where new knowledge supersedes old knowledge in the triumphant march of intellectual advancement.[15] On this model, every story is what Charles Taylor calls a "subtraction story": old ideas are jettisoned when they are *replaced* by new ones. This is the story that academic disciplines in modernity like to tell about themselves. In MacIntyre's view, however, knowledge in fact advances when disciplines function as traditions, where advances in knowledge and understanding are organic developments of a heritage. From the point of view of a tradition like the Christian tradition, "reasons" and "advances" are understood differently

11. Imagine that soccer develops as a game that requires one to touch the ball only with body parts below the waist. But then later it is judged that I can still be playing the same game and perhaps touch the ball with my belly, or even my head. But somehow it is also discerned, internal to the game players, that touching with my hands or elbows is a foul.

12. For an appreciation of this as a Catholic posture, see George Weigel, *Evangelical Catholicism: Deep Reform in the 21st Century Church* (New York: Basic Books, 2014).

13. It should also be noted that such a community of practice engaged in discernment would also need the requisite *virtues* to pull this off.

14. Again, we're thinking of MacIntyre, *Three Rival Versions of Moral Enquiry: Encyclopaedia, Genealogy, Tradition* (Notre Dame: University of Notre Dame Press, 1991).

15. While we are contrasting the "encyclopedic" episteme of the academic guilds with the "tradition-based" reasoning of the church, one could also point out how the "encyclopedic" is science's "tradition." But we're not pressing that point here.

because there is a weight granted to the tradition *as* tradition; there is a requirement that any advance be seen as an extension, not a supersession, of the tradition. There are no prizes for novelty in a tradition.

What's the upshot of all this "meta" framing? It cuts two ways: On the one hand, it should remind us that no tradition worth its salt aims "simply to repeat or paraphrase the tradition."[16] Extension, revision, expansion, and development are intrinsic to a tradition *qua* tradition. We should expect some "modifications" across the heritage of a tradition and should not be surprised if some doctrines are "reformulated."[17] On the other hand, this account helps us to see that any modifications, revisions, and reformulations will (a) need to provide an account of how they are *faithful extensions* of the tradition and (b) have to concede that the discernment of what counts as faithful extension is determined *by the community of practice*, and not just the realm of "expertise." So we will indeed have to determine whether reformulations violate the "core"[18] or "essential" markers of the tradition; and we will have to concede that the determination of this is entrusted to the people of God, which is wider than the realm of academics, scholars, and scientists (though scholars and scientists who are part of this community of practice also get to participate in this discernment process).

Would this mean all is fair game? That everything's up for grabs? That we can revise at will? No, clearly not. Again, the church will have to collectively discern what constitutes a faithful extension of the tradition. Perhaps we might determine that the picture of a historical couple lapsing in a single episode is not essential. But we might also discern that making fallenness basically synonymous with finitude violates the "core" of the traditional doctrine.

Structure of the Volume

The problems raised for the Christian tradition by scientific theories of human origins range across the fields of biology, theology, history, scripture, philosophy, and politics. All of these disciplines are represented in the chapters of this volume.

16. Daniel C. Harlow, "After Adam: Reading Genesis in an Age of Evolutionary Science," *Perspectives on Science and Christian Faith* 62, no. 3 (September 2010): 192.

17. Harlow, "After Adam," 191, 192.

18. John R. Schneider, "Recent Genetic Science and Christian Theology on Human Origins: An 'Aesthetic Supralapsarianism,'" *Perspectives on Science and Christian Faith* 62, no. 3 (September 2010): 197.

In Part I we map the territory of questions and challenges and get a sense of the "lay of the land" at the confluence of science and Christian theology with respect to human origins and the Fall. To open, biologist Darrel Falk provides a comprehensive overview of the "state of the question" from the sciences. He first provides a clear, concise, and comprehensive account of the archeological record about the emergence of *Homo sapiens* and the evidences for common ancestry. This is amplified with a remarkably clear explanation of the genetic evidence for early human populations, including a helpful explanation of how geneticists draw such conclusion. But Falk then presses the conversation in two directions. On the one hand, he emphasizes why Christians need to take such evidences seriously, and shows us *how* to do so faithfully. On the other hand, he pushes back on naturalistic accounts that too quickly conclude to anti-theistic implications of this evidence.

In a second chapter, theologian and ecologist Celia Deane-Drummond stages a conversation between science and theology in light of Roman Catholic teaching on original sin. Surveying the various stances of papal statements and theological proposals, she argues that the engagement between theology and science need not be a one-way street with science dictating the terms to be accepted by theology. Theology, she argues, can also open up new questions for evolutionary theory. She models this by showing how traditional, orthodox theological convictions about human origins and original sin concur with, and even illuminate, recent evolutionary accounts of communal behavior and "niche construction"—even if there are aspects of the traditional doctrine (like biological propagation of original sin) that need to be reconsidered.

Philosopher Jamie Smith then provides an analytic account of what's at stake in debates about the viability of the "traditional" doctrine of the Fall in light of the sort of evolutionary evidences summarized by Falk. Teasing out the intuitions of the Augustinian account of the Fall—an account affirmed by both Roman Catholic and Protestant traditions—he argues that the doctrine of original sin is not just an account of human sinfulness and the need for redemption. Rather, the doctrine of the Fall is integrally an account of the origin or beginning of sin, and such an account is crucial to maintaining the goodness of God. So what "stands" on the Fall is, in fact, not just a matter of theological anthropology but also the doctrine of God. As such, the "historical" or "event-ish" nature of the Fall is crucial to the doctrine. In a closing thought project, he considers how an event-ish understanding of original sin could be imagined as consistent with an evolutionary account of human origins.

The second part of the book then dives down into biblical sources and the traditional theological accounts, mining them as resources for theological imagination. In chapter 4, Old Testament scholar Richard Middleton offers a close reading of the Genesis account in light of evolutionary theory—not positing evolutionary intentions to the authors of Genesis, but rather staging an encounter of mutual illumination. His almost midrashic attention to the layers of allusion and play in the text deepens our understanding of the traditional account of evil and sin while also inviting us to read Genesis anew.

In the next chapter, Joel Green examines the New Testament's contribution to Christian doctrines of original sin and the origins of sin. Green first explores Jewish texts on Adam from the Second Temple period, and then analyzes Paul and James on the character of sin. Green finds that neither set of texts refers to a "Fall" as an event, and neither indicates that humanity's sinfulness is determined by Adam's sin. Green suggests that a careful reading of Paul and James would be amenable to an account of the Fall that would be compatible with scientific evidence, that is, an account of the Fall as a gradual emergence of sin as a pervasive quality of human experience.

Theologian Aaron Riches then mounts a "poetic apology" for the traditional view of sin as issuing from a historical event undertaken by a concrete person. Riches attempts to distance this position both from the view that evolutionary theory is a certainty that relegates Adam to mere myth or metaphor, and from the opposing view that the Bible consists of fragments of data that function on the same plane as the data of science. Riches contends instead that Adam can only be understood within the *figura* of the whole of scripture, which is united by the figure of Christ. The old Adam can only be understood in light of the new Adam, Jesus Christ. Just as Christ is a person, not an abstract idea or metaphor, so Adam must be a concrete person, though one shrouded in mystery, the original protagonist of a history marred by sin that receives its answer in the historical event of Jesus Christ.

In Part III we consider some of the cultural implications of the Fall beyond a narrow consideration of "origins." Ethicist Brent Waters sees at the heart of Christian reflections on the Fall a sense that human life is not as it should be, and a critique of the human impulse to overcome the Fall on our own terms through our own efforts. Waters then analyzes transhumanism—the attempt to overcome human limitations such as aging and death through technology—as the most recent and troubling attempt to perfect humanity through human powers alone. Despite its claims to be purely secular, transhumanism is a type of religion, argues Waters, a heretical mutation of Christian eschatology. Waters critiques the likelihood of social upheaval and the marginalization of those who stand in the way of transhumanist "progress," and he recommends the Christian recognition of our fallen state and our need to forgive and be forgiven.

Norman Wirzba argues that a Christian description of the world *as creation* has important implications for the way we understand the world and our responsibilities within it. Without such an account, the world's fallenness and flourishing become unintelligible. More specifically, Wirzba shows that a Christological narration of creation—a narration clearly begun in scripture, but then developed powerfully by theologians like Irenaeus, Athanasius, and Maximus the Confessor—represents a profound challenge to contemporary accounts of nature as a (sometimes beautiful, sometimes pointless, depending on who is offering the account) realm of unremitting struggle and competition. The teaching of creation is not simply a more or less scientific teaching about the origins of the world. It is also a teaching about the salvation and reconciliation of all creatures and the world with God. Understood this way, the doctrine of creation carries within it important insights that enable us to address questions like the nature of sin, the mission of the church, and the meaning and purpose of human life. When creation is also understood in its eschatological light we can describe fallenness as a creature's inability (or refusal, in the case of humans) to find its fulfillment in God.

In the final section of the book are two historical studies that give us a long view to reconsider contemporary pressures and questions. Bill Cavanaugh's chapter argues that the decline of the Fall narrative in modern thought comes about first for political, not scientific, reasons. The Fall was crucial in medieval political thought for marking the difference between the way the world is and the way the world was meant to be, an eschatological view that destabilized any human claim to power. The Fall disappears in early modern political theory, replaced by a de-eschatologized "state of nature" that justifies political power as a response to the way things simply are. Through the figures of Hobbes, Filmer, and Locke, Cavanaugh shows how the "naturalization" of the Fall in early modern political theory contributes both to the rise of the modern state and to the divorce between theology and political science and between theology and natural science. Both natural science and political science lose teleological and eschatological reference, but Cavanaugh argues that it need not be the case. If we can see that the divorce of theology and science in the West has been promoted by nonscientific, political, factors, then perhaps we can see that antagonism between science and theology is by no means inevitable.

Peter Harrison's contribution examines the way we think about conflict between religion and science. Because we are so familiar with instances of "bad" conflict between science and religion—such as the Galileo trial and the simple rejection of evolution by some Christians—we assume that conflict between religion and science is always to be avoided. Harrison argues, however, that there are instances of "good" or justifiable conflict or tension between science and religion, and that Christians should not too hastily assume that Christian doctrine should immediately be adjusted to suit whatever is the reigning scientific paradigm. Harrison argues that conflict between science and religion is never inevitable but is always possible, depending on what the current claims of each are. Each potential science-religion conflict needs to be considered on a case-by-case basis, and the general claims of a scientific theory—in the case of evolution, descent with modification—need to be distinguished from the specific mechanisms and implications of the theory, which tend to be much less certain. Harrison draws on the thought of Augustine to exemplify a proper Christian approach, and examines some historical cases in which conflict was, from a Christian point of view, justified.

Our hope is that, collectively, these contributions will help us think carefully about the stakes and parameters of these conversations, while also modeling ways to confirm that in Christ "all things hold together."

PART I

Mapping the Questions

1 Human Origins

The Scientific Story

DARREL R. FALK

Scientific investigation portrays a picture of human origin that is gradual, not instantaneous, occurring through an evolutionary process that is frequently depicted by scientific pundits as atheistic. Christians, however, unanimously disagree with the latter, but express their disagreement in different ways. Some articulate their disagreement by declaring the science fundamentally flawed. These Christians seek a whole new way of doing the science of human origin. The problem with this approach is that it is pitting Christianity against a vast expanse of widely tested scientific data and likely doing so for reasons that are theologically unnecessary. Certainly, if the scientific paradigm of our evolutionary origin is correct, it raises all-important questions about the nature of the Fall and the origin of sin, issues that are addressed in this volume. We can address those questions with a spirit of optimism, however, because if creation has taken place through a gradual process rather than an instantaneous one, and if the foundational propositions of Christianity are true, then the traditional understanding of theological precepts will only be enriched as they are explored in this new light. But what is that new light? What precisely do the scientific data have to say about how we got here? Answering that question is the purpose of this chapter.

Tracking Fossils

The fossil record of our emergence from the great ape lineage is found in Africa. Here a variety of fossilized skeletal remains from many transitional species have been found that illustrate features of the change from walking on four legs to two, from using forelimbs to swing from trees to using them for the finely tuned manipulation of objects, from small brains to large, and from the face of an ape to the facial features of a human being.

It is one thing, however, to identify transitional features in the fossil record, but it is another to show that their existence was timed in a manner consistent with a progressive sequence from ape to human. Nowhere is the progressive timing of the transitions better documented than in the Great Rift Valley of northeastern Africa where Ethiopia, Kenya, and Tanzania exist today. This valley has come into existence through the geological splitting action caused by two continental plates that have been sliding apart from one another for millions of years. The resulting valley is still subject to flooding just as it has been throughout this span, and animals still get stuck in the muddy flood sediment just as they have for millions of years. After their bodies decay within that mud, the resulting skeletal remains—if undisturbed—are cemented in place. As the sediment continues to harden over future millennia, they will become tomorrow's fossils.

The age of fossils can be readily determined because of another unique geological feature of the region. The geological instability caused by the sliding continental plates results in frequent volcanic eruptions, and this sporadic activity produces precise layers of ash embedded in the sediment just like the fossils. So although the fossils can't be dated (they are too old for C^{14} dating, which has a limit of about 50,000 years), the ash can.

The 4.4 million-year-old species, *Ardipithecus ramidus*, was identified in this way. In November 1994, a single handbone was found peeking out of some sedimentary rock. Careful excavation of the site resulted in the recovery of 45 percent of the skeletal remains from one individual, a female that came to be known as Ardi. It is likely that Ardi was bipedal like us. She certainly didn't have the skeletal features that would have enabled her to be a knuckle-walker like chimpanzees and gorillas. Instead, she had features that pointed to a lifestyle adapted to both trees and ground.[1] Like today's great apes, she had a splayed big toe sticking out sideways that would have served well in tree climbing. She had hands and arms that were well suited for life in the trees as well. There is no sign of stone tools at any of the many archeological sites of this age or even up to a million years younger, so it is unlikely that members of her species were using tools.

---

1. C. Owen Lovejoy et al., "Combining Prehension and Propulsion: The Foot of *Ardipithecus ramidus*," *Science* 326 (2009): 72.

As investigators move forward in time, the fossil finds become characteristically different. *Australopithecus afarensis*, a species of which the famous Lucy was a member, is one example, but there are other fairly complete specimens identified as well. The best evidence for their bipedality is a set of 3.6 million-year-old footprints of two individuals extending over a distance of about eighty feet in what would have been wet volcanic ash. The big toe was parallel to the other toes; it was not splayed outward as with Ardi. Detailed analysis of the gait of the pair indicates that they walked in a manner almost indistinguishable from us. The footprints likely belonged to Lucy's species; indeed *A. afarensis* fossils were found nearby in the same layer of volcanic ash. Detailed analysis of the anatomical characteristics shows that the shinbone, for example, was structured such that it would have attached to the ankle in a manner that resembled how ours is attached rather than the way it is joined in the chimpanzee. Moreover, the structure of the ankle itself is human-like. On the other hand, the face was apelike with a flattened nose and strongly protruding lower jaw. The brain was about one-third the size of ours. The shoulder blade was not humanlike; it resembled that of a gorilla. In short, *A. afarensis* had a body that would have been well adapted for an existence in trees as well as on the ground. Interestingly, the hyoid bone, a part of the voice box, was structured in a manner that is much closer to that of a gorilla than a human, and this suggests Lucy and her kin had ape-like vocal abilities.[2] Multiple other species of the genus *Australopithecus* have been found in both this region and southern Africa at locations dated as recently as 2 million years of age—1.5 million years after the earliest members of the genus. Some of these were undoubtedly cousin species not on the direct lineage to *Homo*—side branches on the species-tree that died out without contributing to our own lineage.

Beginning at sites about 2 million years of age, fossils that share an increased number of our characteristics appear on the scene. The individuals from whom those fossils are derived had a large brain case and a narrower, less stocky body shape. Similarly, their arms were shorter and legs longer. Our genus, *Homo*, had emerged. Several species within the genus have been identified, but one in particular is especially widespread in the fossil record. *Homo erectus* apparently inhabited the region beginning about 1.9 million years ago. Within 100,000 years members of the species apparently made the journey into Asia. Fossil remains that are 1.8 million years old have been

2. Ian Tattersall, *The Strange Case of the Rickey Cossack and Other Cautionary Tales from Human Evolution* (New York: Palgrave Macmillan, 2015).

found in a cave in the Republic of Georgia, and 1.6 million-year-old remains have been found near Beijing in China as well as Indonesia. Indeed, the species persisted in parts of Asia until as late as a few hundred thousand years ago.[3] During the long life of this species, the brain size was increasing. The earliest skulls found would have had room for brains only slightly larger than those of apes. However, over time there was a gradual increase in brain volume until it eventually reached a size close to that of our own.[4] So the geographic distribution of *Homo erectus*—from various locations in Africa to many parts of Asia—was widespread, although we know few other details.

In September 2015, the largest and most complete find ever for any hominin[5] species, *Homo naledi*, was announced.[6] This species, found in a South African cave, bears some resemblance to *Homo erectus* with its small brain, anciently structured pelvis, shoulder, and ribcage. However, its hand, wrist, foot, and ankle are quite similar to our own. At this writing, the specimens remain undated.

It is in the Rift Valley that our own species, *Homo sapiens*, first appears in the fossil record. Those first skeletal remains are 195,000 years old. Preceding this, another species, *Homo heidelbergensis*, lived from about 700,000 to 200,000 years ago, and many investigators consider it to have been an ancestral species of our own. *H. heidelbergensis* is represented in the fossil record as far south as South Africa, as far north as England and Germany, and as far east as China. Indeed, as a result of work just published in 2015, we now have a significant amount of the DNA coding information from one set of 300,000-year-old *H. heidelbergensis* bones.[7]

Although hints of this story began to emerge late in the nineteenth and early in the twentieth century, most aspects of it have only become fully apparent in the past forty to fifty years. Thus, until just the last few decades, detailed knowledge of our history—how we came to be who we are—only stretched back several thousand years. Suddenly though, as a result of this

---

3. Daniel E. Lieberman, *The Story of the Human Body: Evolution, Health, and Disease* (New York: Pantheon Books, 2013).

4. For details see P. Thomas Schoenemann, "Hominid Brain Evolution," in *A Companion to Paleoanthropology*, ed. D. R. Begun (Chichester, UK: Wiley-Blackwell, 2013), 136–64.

5. The term "hominin" refers to any species within the lineage that includes ours, *Homo sapiens*, but not that of the great apes. Some of the literature still uses the older term, "hominid."

6. L. Berger et al., "*Homo naledi*, a new species of the genus *Homo* from the Dinaledi Chamber, South Africa," *eLife* 4 (2015): e09560.

7. Ann Gibbons, "Humanity's long, lonely road," *Science* 349 (2015): 1270.

knowledge-explosion, we now can go back millions. We have the bones—the skeletal remains of our ancestors. We can see when and how their anatomy changed, becoming more and more like our own. And we can see that finally about 200,000 years ago their skeletal features became indistinguishable from ours.

Beginning about the same time as modern humans (*H. sapiens*) appear in the fossil record in the Great Rift Valley region of Africa, the Neanderthals (*Homo neanderthalensis*) emerge in the fossil record at various locations in Europe and western Asia. Based on the many artifacts found at Neanderthal archeological sites, it is clear that they were skilled and intelligent hunters, equipped to survive the cold arctic conditions of the ice age. However, they showed little sign of creative activity. Their stone-working tools did not vary much over the more than 150,000 years of their existence. Neanderthals would sometimes adapt old tools to new uses, but unlike *Homo sapiens* (see below) they did not excel at inventing new technologies. This is a major difference between *H. sapiens* and *H. neanderthalensis*.[8] Neanderthals had a long, low cranium, a huge face, a large nose, marked brow ridges, and no chin. Their brain, perhaps because of their brawny physique, was about 10 percent larger than ours.[9] The Neanderthal fossil record ceases abruptly at 39,000 YBP (years before present). What happened to them? Most investigators think that it is not just a coincidence that they vanished from the earth only several thousand years after our species appeared in what had been their region. We have a long history of causing extinctions of other species whenever we move into a new world area or expand our technology and influence. This pattern of destruction—the extinction of other species—likely began tens of millennia ago as we conquered a new land already occupied by our cousin species, the Neanderthals. It seems that our own species came bursting onto the Neanderthals' scene and in some unknown way brought about their demise, after they had been living in harmony with their environment for tens of thousands of years.

To summarize, it appears that our species was confined to Africa for all of its early history. However, beginning about 100,000 years ago that changed. Recent fossil discoveries in southeast China and Israel show that *H. sapiens* had forayed out of Africa; although we can't be sure their lineages in these locations persisted.[10] The oldest *Homo sapiens* fossils in Australia

8. Tattersall, *The Strange Case of the Rickey Cossack*.
9. Lieberman, *Story of the Human Body*.
10. Ann Gibbons, "First modern humans in China," *Science* 350 (2015): 264; Chris

date to about 60,000 YBP, those in Europe are about 42,000 years old,[11] and in the New World they date back at least 12,500 years,[12] although stone tools were recently found in a Chile settlement dated to about 18,000 years ago.[13]

Tracking Genes

Although the skeletal remains poignantly tell a portion of the story of the origin of our species, the story they tell is limited by the fact that they are inert—each set of fossilized remains is a single snapshot of a distant past. Besides the bones, though, we also have another physical component—our genes—a remnant of our ancestors' existence that is not inert and still lives on in each of us. In contrast to the fossils, our genes provide, not a snapshot, but a moving story of their past. And their story is an important part of our story.

The molecule that houses our genes is DNA. Inherited from each of our parents, DNA has the instructions for building the human body. It serves as a code. There are 6 billion units (called bases) in the code—3 billion from each parent. DNA has four code "letters," the well-characterized molecular components A (adenine), G (guanine), C (cytosine), and T (thymine) arranged in a specific sequence that is read and interpreted by the cellular machinery. The code mutates (i.e., changes) ever so slightly from one generation to the next. We can measure the rate of change precisely, and an average of sixty mutations (in the 6 billion code letters) occur each generation. As we go further back in time, this means there are more and more DNA differences between our ancestors and us. For example, of the sequence of 6 billion code letters there would be 120 alterations not found in any of your four grandparents, and going back even further, 180 found in you but not your eight great grandparents. If common descent is true, then one can calculate how many mutations would have occurred in the 7 million years or so since the

Stringer, *Lone Survivors: How We Came to Be the Only Humans on Earth* (New York: Times Books, 2012), 46.

11. Stringer, *Lone Survivors*, 46.

12. Morten Rasmussen et al., "The genome of a Late Pleistocene human from a Clovis burial site in western Montana," *Nature* 506 (2014): 225–29.

13. Ann Gibbons, "Oldest stone tools in the Americas claimed in Chile," *Science* (November 2015). For a detailed account of the impact of our forays into these new lands, see the very important book by Elizabeth Kolbert, *The Sixth Extinction: An Unnatural History* (New York: Henry Holt, 2014).

common ancestral species of both chimpanzees and humans existed. When that calculation is done, it turns out that the number of expected mutations corresponds within a factor of two to the actual number of coding differences between the two species. To be that close to the predicted value after 7 million years of passing the DNA from parents to offspring (about 350,000 generations) is truly amazing.

However, its significance in demonstrating our evolutionary origin is substantiated even further by examining the frequency of a particular type of mutation. The code letter C mutates to a T about eighteen times more frequently if that C is adjacent to a G than is the case if it is adjacent to any of the other three code letters combined. If the genetic difference between humans and chimpanzees were caused by mutations in each of the two lineages, then the sequence in the code would be altered eighteen times more frequently in cases where a C had been adjacent to a G. This is exactly what is observed.[14] Since the solidity of this evidence may seem a little esoteric, here is an illustration of what it means. Let's say there are two similar written stories in manuscripts found in each of two neighboring ancient communities. You are interested in whether each story was independently written or if they arise from parallel copying from a single manuscript. You look closely and you find that where the two manuscripts differ most is at positions where letters of the alphabet are easily confused if the scribe's writing is sloppy. For parts of the alphabet where the letters are highly distinctive and seldom confused, the two manuscripts are almost identical. The observation that manuscript-differences are correlated with susceptibility to copying error convinces you beyond doubt that there was a single manuscript that has been independently copied by scribes representing each of the two communities.

The chimpanzee and human genomes are two manuscripts that differ slightly. In places where we know from test tube experiments that copying errors occur frequently, they differ eighteen times as often compared to more stable sites. This, when coupled to the overall mutation rate data described above, provides overwhelming evidence for a single ancestral species. The single original "manuscript" has been copied independently through two separate lineages for millions of years.

There are many other ways in which the tracking of DNA changes demonstrates common ancestry between humans and the great apes, and the results are fully consistent with the fossil data that traces the actual an-

---

14. Mark Jobling et al., *Human Evolutionary Genetics*, 2nd ed. (New York: Garland Science, 2014), 54.

atomical changes. This is thoroughly discussed in a very fine book by the Christian geneticist Graeme Finlay.[15]

Given the knowledge of how often mutations happen, geneticists are able to examine details about many other aspects of the history of our species. For example, genetic analysis demonstrates quite clearly that all whose traceable ancestry is non-African are descended from about one thousand people who left Africa approximately 50,000 to 70,000 years ago.[16] Although there is no reason to assume these ancestors migrated out as a single group or that they left within a single generation, what is well established is that all Europeans, Asians, and aboriginal peoples of Australia, the Pacific Islands, and the New World are descended from this relatively small group in somewhat recent prehistory.

How do we know this? Consider the following analogy in which we will think back once again to the period before the existence of printing presses. In that era, of course, scribes prepared multiple copies of manuscripts. For the sake of illustration, let's say that a single scribe is assigned the task of copying a one-page manuscript. Through a long and arduous process he prepares one thousand copies. He does so, not using the original as the template for each new copy, but instead using the most recent copy as his template for the next one as he works through the task of completing one thousand copies. No scribe is perfect of course. Let's say the error rate is one per one hundred copies, so by the time he gets to #1000, there are ten errors in the manuscript. Finished with his task, the scribe sends all except the last one off to a warehouse for storage, but keeps the most recent one with him. Sadly, there is a fire in the warehouse, and the earlier 999 are all destroyed leaving only the last one with its ten errors. At that point he receives a request for twenty-five. He quickly copies twenty-four more. Each of the twenty-five that he passes on has the same ten errors. There is a total of ten "mutations"; each of the twenty-five manuscripts carries the same ten.

---

15. Graeme Finlay, *Human Evolution: Genes, Genealogies and Phylogenies* (Cambridge: Cambridge University Press, 2013).

16. The fossil data indicate that our species may have migrated out of Africa as long as 100,000 years ago, so this genetic data, although roughly comparable, is nonetheless different than predicted by the fossil record. It is possible that the earlier populations died out without contributing to our lineage. Alternatively, one or both sets of dating mechanisms may not be sufficiently accurate. The fact that two totally different mechanisms (genetics and radioisotope dating) are within 1.5 to 2 fold of each other is actually pretty strong confirmation of the reliability (within limited error) of the dating mechanisms. See Eugene E. Harris, *Ancestors in Our Genome: The New Science of Human Evolution* (New York: Oxford University Press, 2015).

Now let's examine another scenario. Let's say that there are twenty-five scribes doing the replicating—each copying the most recent reproduction as his template until each gets to his #1000. The twenty-five all send their earlier 999 copies to the warehouse, keeping the final one (#1000) with them. The warehouse burns down and destroys all except the twenty-five—the one that each of the twenty-five scribes has kept with him. When the request for twenty-five comes in, they each pass along their one final copy. Each of the twenty-five has about ten errors, but in this case for the sample of twenty-five manuscripts the errors are different, making for a total of ten times twenty-five errors—two hundred and fifty in all. In the first scenario, the diversity in the manuscripts is much lower—only ten errors in the whole group of twenty-five manuscripts. The cause of the increased "diversity" in the second group of manuscripts (two hundred and fifty errors versus ten) is the "population size" of the scribes.

So in summary, if we know the error rate (one in one hundred in this case) the number of copying events (one thousand), and the number of errors, we would be able to determine the average number of scribes. That, in essence, is how mean historical population size can be estimated from the amount of genetic diversity in the human population.

This, however, is only half of what we need to consider. Unlike the constant number of scribes over time in the above example, human population size is not constant. In the out-of-Africa scenario, geneticists estimated the population size at particular times and determined that there was one specific time period when the population was especially small. What is the rationale behind this calculation? There is, in essence, a "ticking clock" built into human chromosomes that we can use to tell the population size at any given time in ancient history. DNA consists of long stretches of contiguous versions of code that remains connected in long blocks. As time goes by the blocks get shorter through a process known as recombination. The rate of shortening has been carefully worked out. Given this, it is possible to determine how many new mutations occur in the human population at specific time points. The number of mutations generated at a given time, as we saw with the "population size" of copying scribes, is directly correlated with the number of individuals in whom the DNA replication is taking place. It is in this way that geneticists have estimated that about 50,000 to 70,000 years ago there was a population "bottleneck," and all of today's non-Africans are descended from the relatively small number of individuals that left Africa at about that time.

We now turn to the human population as a whole. The genetic diversity of humanity is much greater when we consider both Africans and non-Africans. Geneticists are agreed that in Africa there was never a time when the population size of reproducing individuals was less than 10,000. Indeed, beyond that, through all of the past seven million years of hominin history, our lineage (those who contributed to our gene pool) was never much larger than 10,000.[17]

Much ado has been made in certain Christian circles about the scientific evidence for a mitochondrial Eve and a Y-chromosome Adam, as though science really does point to a single pair as the genetic progenitors of the entire human race. This is not the case, and the scientists who chose those names for the phenomenon they were describing have likely long since regretted their name-choice. It is true that there was a time when a particular individual lived who had the Y chromosomal DNA that is ancestral to that carried by all human males. This male lived about 240,000 years ago. This does not mean that there were no other males alive at the time; indeed as just mentioned there were thousands. What it *does* mean is that any other versions of the Y chromosomes present in the human population at that time are not present today. Similarly, the DNA in the mitochondria of every person alive today is derived from that present in a single woman who lived about 165,000 years ago. (The mitochondria are the energy-generating compartments of a cell. They contain DNA, and that DNA is inherited exclusively through the eggs of mothers.) For various reasons, mitochondrial DNA and the Y chromosome were the easiest to study in the early days of genomic research. We now can study the whole genome. From these studies, we know that each of the 21,000 human genes was once present in one particular person a long time ago. However, the individual who carried that one version for each specific gene is different than the individuals who carried the version of the other 21,000 genes that all of us have. Each gene in the human population was once found in only one person: 21,000 genes and 21,000 different individuals. Not only that, but the time when each of the various individuals who had each ancestral gene lived is widely different from gene to gene. In some cases, it was as short as 100,000 years ago; for other genes it was as long as several million years ago. It does not in any way mean that there were no other people alive at the time; it is simply a matter of something that happens over long periods of time when there are only 10,000 or so individuals contributing to the gene pool. Speaking personally, even today, my specific version of the Y chromosome is lost forever from

17. Harris, *Ancestors in Our Genome.*

the gene pool. I have two daughters and no sons. My Y chromosome with its long heritage dating back millions of years has at last reached its Waterloo. The same thing has been happening down through ancient history with versions of genes that may or may not get passed on—especially when the population size is only 10,000 individuals for hundreds of thousands of years.

Tracking the Origin of Human Uniqueness

Archeological evidence hints that about 100,000 years ago there was a dramatic change in human evolution—*Homo sapiens* had acquired the ability to think in a manner that included the use of symbols to describe their perception of reality. Language is a form by which objects are represented symbolically. Art and spiritual expression also require forms of abstract (symbolic) thinking. Investigators do not know the specific anatomical and physiological details or the sequence of events associated with how this came about, but it is clear this trait, symbolic cognition, set us on a journey that revolutionized our world. Indeed, as Ian Tattersall summarizes the findings, "it seems altogether reasonable to believe that [symbolic cognition] was acquired as an integral part of the larger developmental reorganization that gave rise to the distinctive modern body form."[18] In his assessment of the origin of this shift, Tattersall goes on to state:

> Given that evolution always has to build on what was there before, it seems reasonable to conclude that modern humans achieved their cognitive uniqueness by grafting the symbolic faculty onto a preexisting high intelligence of the ancestral intuitive kind—possibly exemplified by that of *H. neanderthalensis*—rather than by replacing that older style of intelligence wholesale.[19]

In the preceding 99 percent of hominin existence, symbolic thinking did not apparently exist. What is significant about that which emerged after the origin of our species was not increased skill at carrying out complex tasks—our predecessors had indulged in complex activities like the making of sophisticated tools—what became cognitively unique in humans was that ability to symbolize. Evidence for this includes carvings with designs

18. Tattersall, *The Strange Case of the Rickey Cossack*, loc. 3634.
19. Tattersall, *The Strange Case of the Rickey Cossack*, loc. 3652.

conveying meaning, and the use of ochre in new ways indicative of abstract thinking. This is what appears almost instantaneously as the last one percent of hominin existence begins. Indeed so sudden is this appearance that, as Tattersall puts it, "there is no way our unusual cognitive capacities can be the perfected products of long-term selective pressures."[20] By this, Tattersall does not mean that he thinks something supernatural happened in human evolution. What he means is that our brains evolved over a long period of time for something other than the use of symbols. However, once that "something" happened, our species moved into an explosion of ever-accelerating creativity that continues at an ever-increasing pace in today's world.

What was that "something" that changed everything? Tattersall argues that it was language:

> Like symbolic thought, language involves mentally creating symbols and reshuffling them according to rules; so close are these two things that it is virtually impossible for us today to imagine one in the absence of the other. What is more, it is relatively easy to envisage, at least in principle, how the spontaneous invention of language in some form could have started those symbols chasing around early modern human minds in a structured way. By the same token, it is no problem to understand how language and its cognitive correlates would rapidly have spread among members, and ultimately populations, of a species that was already biologically enabled for them.[21]

Yuval Harari summarizes the uniqueness of human language this way:

> A green monkey can yell to its comrades, "Careful! A lion!" But a modern human can tell her friends that this morning, near the bend in the river, she saw a lion tracking a herd of bison. She can then describe the exact location, including the different paths leading to the area. With this information, the members of her band can put their heads together and discuss whether they should approach the river, chase away the lion and hunt the bison.[22]

20. Tattersall, *The Strange Case of the Rickey Cossack*, loc. 3712.
21. Tattersall, *The Strange Case of the Rickey Cossack*, loc. 3689.
22. Yuval Noah Harari, *Sapiens: A Brief History of Humankind* (New York: Harper, 2015), 22.

Some investigators place at least as much emphasis on the significance of another uniquely human quality—the development of a full Theory of Mind (ToM). Ajit Varki and Danny Brower outline this in their book, *Denial: Self-Deception, False Beliefs, and the Origin of the Human Mind*.[23] A full ToM is the realization that another individual has an independent mind like his or her own. It is the full awareness of the self-awareness of others. No other species, they and others argue, has this ability.[24] Irrespective of whether the development of a full ToM happened suddenly as Varki and Brower argue, or more gradually to facilitate cooperation in hunter/gatherer societies,[25] it is clearly essential for many of the uniquely human attributes. Varki and Brower list a whole series of human traits that are dependent on a full ToM: active care for the infirm, concern for posthumous reputation, death rituals, food preparation for others, "grandmothering" (including doting), healing of the sick, hospitality, rules of inheritance, the concept of justice and laws governing it, storytelling, multi-instrumental music, religiosity, teaching, and the act of torture designed to break the spirit of another.[26]

Regardless of the main contributing factors, it is clear that a sort of evolutionary big bang of cultural innovation with many accompanying changes began uniquely in our history about 100,000 years ago when most, if not all, humans were still in Africa.

"Historical Contingency" and the Origin of Our Species

Throughout this book, there is a clear assumption that humankind and all of nature are here by divine decree and they continue to exist because of the Creator's ongoing presence. "He is before all things and in him all things hold together," we are told in Colossians 1:17 (New Revised Standard Version). Given that this is the presupposition upon which the book is based, it is important to point out that there is complete discordance between this premise and the assumption that pervades the writing of biology's most renowned spokespersons. The clash is not so much with the pundits' argument that natural selection is an impersonal force giving rise to life's diversity. Many

23. Ajit Varki and Danny Brower, *Denial: Self-Deception, False Beliefs, and the Origin of the Human Mind* (New York: Twelve, 2014).

24. Michael S. Gazzaniga, *Who's in Charge? Free Will and the Science of the Brain* (Mt. Pleasant, TX: Echo, 2012), 160.

25. Lieberman, *Story of the Human Body*, loc. 2388.

26. Varki and Brower, *Denial*, 103.

Christians would argue that natural selection is fully compatible with theism. It is God's process; put into place by the Creator's foresight, they would say. So it is not natural selection as an impersonal force that is necessarily at odds with the great creation passages in scripture. Rather, the finding is now well established that natural selection alone cannot explain our presence here. What biology has shown most clearly over the past couple of decades is that historical contingency, which the pundits define as pure unadulterated luck, is by far the most crucial component in humankind's arrival as a species. For example, one of the most highly respected biologists of the last sixty years, E. O. Wilson, puts it this way:

> That the human line made it all the way to *Homo sapiens* was the result of our unique opportunity combined with extra-ordinarily good luck. The odds opposing it were immense. Had any one of the populations directly on the path to the modern species suffered extinction during the past 6 million years . . . another 100 million years might have been required for a second human-level species to appear.[27]

Although it is impossible to calculate the probability of an event that has happened only once in earth's history, the view that we are here by luck and not by divine decree is almost unanimous in the minds of evolutionary biologists. Although it is true that biologists agree that once living systems originated, certain biochemical, physiological, and even anatomical pathways become almost inevitable,[28] this does not mean, however, that a particular species or even a particular family of species is inevitable. Consider, for example, the adaptive radiation of mammalian diversity that began to flourish 65 million years ago with the end of the dinosaur era. Although mammals had been present for about 100 million years prior to the extinction of dinosaurs, the explosive radiation of mammalian diversity did not occur until immediately after the event.[29] It is now well established that the dinosaur extinction was caused by the arrival of a six-mile-wide aster-

27. E. O. Wilson, *The Meaning of Human Existence* (New York: Liveright Publishing, 2014), loc. 944.

28. Andreas Wagner *Arrival of the Fittest: Solving Evolution's Greatest Puzzle* (New York: Current, 2014); Simon Conway Morris, *The Runes of Evolution: How the Universe Became Self-Aware* (West Conshohocken, PA: Templeton, 2015).

29. Mario dos Reis et al., "Phylogenomic datasets provide both precision and accuracy in estimating the timescale of placental mammal phylogeny," *Proceedings of the Royal Society of London B* 279 (2012): 3491–3500.

oid traveling about 50,000 miles per hour as it hit earth, creating the most powerful earthquake of all time. The crater has been found; the iridium it carried with it in high abundance has been identified as a distinct layer in the earth's rocks at the expected time; and a layer of minerals formed only through high impact are distinctively distributed in the same layer. The impact triggered a cataclysm of volcanic eruptions, wildfires, tsunamis, acid rain, and sunlight-blocking dust. Still, even this would likely not have been sufficient for the demise of all dinosaurs were it not for one other thing. It is now clear that the dinosaur lineage, especially that of herbivorous dinosaurs, was in the midst of an ecological crisis that created much instability in the food web immediately before the asteroid's arrival.[30] Both events—together exceedingly improbable—provided the opening needed for the mammalian radiation that soon took over the planet. Without those events and likely millions of other smaller events (see below), the lineage that gave rise to us would never have happened.

In fact, the late Stephen Jay Gould put it this way:

> We came this close (put your thumb about a millimeter away from your index finger), thousands and thousands of times, to erasure by the veering of history down another sensible channel. Replay the tape a million times from a Burgess beginning, and I doubt that anything like *Homo sapiens* would ever evolve again. It is, indeed, a wonderful life.[31]

Although Gould is adamant about the very high likelihood that similar biochemical, physiological, and anatomical adaptations to specific ecological niches will occur over and over again (indeed he is famous for this through his classic essay, "The Panda's Thumb"),[32] when it comes to the origin of a particular species like humans, he is just as adamant about their *un*likelihood:

> Evolving life must experience a vast range of possibilities, based on environmental histories so unpredictable that no realized route—the pathway to consciousness in the form of Homo sapiens or Little Green Men, for example—can be construed as a highway to heaven, but must be viewed as a tortuous track rutted with uncountable obstacles and festooned with

30. Stephen L. Brusatte, "What Killed the Dinosaurs?" *Scientific American* 313 (2015): 6.

31. Stephen Jay Gould, *Wonderful Life: The Burgess Shale and the Nature of History* (New York: W. W. Norton, 1989), 289.

32. Stephen Jay Gould, *The Panda's Thumb: More Reflections in Natural History* (New York: W. W. Norton, 1981).

innumerable alternative branches. Any reasonably precise repetition of our earthly route on another planet therefore becomes wildly improbable even in a trillion cases.[33]

Despite the skepticism of noted Cambridge paleontologist Simon Conway Morris about this,[34] it is Gould and Wilson who represent the view of almost all evolutionary biologists. Indeed their conclusions—based on what evolution did and did not produce over millions of years—are highly epistemic. Historical contingency and its role in life's history has been tested over and over again. Here are a few examples, which I present because it is important for theologians, pastors, and an informed lay audience—the primary audience of this book—to grasp the significance of what science does and does not say about the origin of our species.[35]

The lineages of old and new world monkeys have been going their separate ways for about 35 to 38 million years. Today, there are 124 species of new world monkeys, each with distinctive properties such as a flat nose, nostrils that point to the side, and frequently a long prehensile tail that can be a sort of fifth limb used for swinging through the trees. Old world monkeys, on the other hand, have a narrower snout and nostrils that point down. Many have only a short tail or no tail at all.

The first fossils of new world monkeys date to about 35 million years ago, and they bear distinctive structural features also found in an early African species that lived at about the same time. Paleontologists think that both the old and new world monkeys are closely related and that they emerged as distinctive lineages from a single ancestral species. The 35-million-year-old specimens in both Africa and South America are so similar because at that point in time, the lineage-separation had just recently occurred. But how did the two groups become physically separated? Africa, of course, was at one time physically contiguous with South America. The continental split and subsequent separation, however, took place 100 million years ago—more than 60 million years before the branch point in the monkey lineage. So how did they get to the new world? There is almost unanimous consensus now that something like the following would have taken place: a small number (perhaps even a single pregnant female) was trapped on a huge tropical tree

33. Stephen Jay Gould, "The War of the Worldviews," in *Leonardo's Mountain of Clams and the Diet of Worms* (New York: Harmony Books, 1998), 351.

34. Conway Morris, *Runes of Evolution*.

35. See Alan de Queiroz, *The Monkey's Voyage: How Improbable Journeys Shaped Life's History* (New York: Basic Books, 2014), for an excellent analysis of that which follows.

as it floated down river (possibly in a massive flood) and then, having been transported in an ocean current, the tree with its clinging cargo arrived a couple of weeks later on the shores of South America. Once that happened, the rest is history: the next 35 to 40 million years became an interesting experiment in evolutionary biology. What took place in that lineage over that vast landscape with its millions of square miles of widely diverse habitats and perhaps ten million generations? The lineage gave rise to more monkey species—lots of them—but it never gave rise to anything remotely resembling the ape or hominin lineage. Based on the similar fossil findings in both South America and Africa 35 million years ago, on each continent the ancestral species at the start was almost the same, but it was only in Africa that it led to both apes and hominins—besides of course the vast array of old world monkeys that today includes seventy-eight species.

Just as continental drift led to the breakup of Africa and South America, similarly Australia separated from South America about 60 million years ago. Early mammals were "on board," and their progeny in the coming millennia—a group of nonplacental mammals—went on to become the mammalian lineage in Australia. In essence then, this is another experiment in evolutionary biology. What emerged from that 60-million-year-long "test run" on this island continent? Nothing like humans, apes, monkeys or even a new placental mammal body plan ever appeared there—only more mammals with pouches, each beautifully adapted in highly specific ways to all sorts of habitats, as well as a small number of egg-laying species.[36]

Another evolutionary experiment took place on the land mass we now call New Zealand. New Zealand broke away from the large supercontinent even earlier than Australia—about 80 million years ago. It carried no mammals at the time of its breakaway and no mammals ever arrived on floating trees. So the only large animals that ever lived on New Zealand were birds and reptiles. Over a period of 80 million years and a geographical area of 100,000 square miles with a vast array of different climate and ecological niches, nothing with the mammalian body plan ever arose there and certainly nothing like monkeys, apes, or humans.

Madagascar provides more data to consider. With a land area over twice that of New Zealand, it separated from Africa about 130 million years ago

---

36. The platypus and spiny anteater have traits that are derived from an even more ancient form of mammal. All other Australian mammals are marsupials and have pouches, except bats. Australian bat species of course are an exception, which, given their wings, proves the rule.

and from India close to 80 million years ago. Around 54 million years ago a small number of lemurs made it across the 250 miles from Africa to land in Madagascar. Lemurs are primates like us, and we would have shared a common ancestor with them about 60 million years ago. On Madagascar, lemur evolution thrived and eventually gave rise to over 100 different species of lemurs. Despite 54 million years and a very substantial tropical land area with diverse ecosystems and a primate ancestral species, this experiment also yielded no monkeys, apes, or hominins.

Even in Asia and Europe with their vast land area and with plenty of primates present throughout the past 65 million years, nothing like hominins ever originated there. True, *Homo erectus* migrated to there from Africa two million years ago and a number of other *Homo* species, like Neanderthals, likely originated there from within the hominin lineage, but the hominin lineage itself began in Africa and nothing like it originated anywhere else, not even in Asia or Europe.

In fact, based on recent data, it is beginning to appear that the origin of the anthropoid lineage itself (monkeys, apes, and hominins) in Africa got its start with an extremely unlikely event in its own right: the journey of a few individuals of an ancestral species from Asia to Africa across a large ancient body of water called the Tethys Sea. At the time, about 40 million years ago, the supercontinent was still breaking up and the Tethys Sea separated Asia and Africa. The fossil remains of an early primate with characteristics that make it a likely ancestor[37] to all Anthropoids has been identified in what is now Egypt. Interestingly, this species had a closely related cousin species thousands of miles away across the sea in Myanmar (Burma). To have two similar species present at about the same time thousands of miles apart but separated by a great sea, suggests a scenario not unlike that which we met earlier for Madagascar's lemurs and the new world's monkeys. There is evidence that the lineage started in Asia, so the direction of this sea-bound journey would have been Asia→Tethys Sea→Africa. Were it not for this unlikely trip of a few organisms (perhaps just one pregnant female for all we know) to Africa, a land with a whole new set of ecological niches, it seems likely that no monkeys, apes, nor the lineage to humankind itself would have ever taken place. That one "accident," may well have been responsible for our entire lineage. Without it, Gould, Wilson, and most leading biologists would likely say humans would not have come into existence.

---

37. Or a species closely related to the one that actually made the journey.

"Historical Contingency" and the Providence of God

This notion that natural selection, if given a chance—indeed, if given millions of opportunities by rerunning the tape of life over and over again—would almost certainly not produce anything like us, is likely the most theologically significant proposal that emerges from the biological sciences today. However, it is based upon a premise that is philosophical, not scientific. It is firmly grounded in the proposition that there is no such thing as divine providence.

When biologist Henry Gee summarizes our origin by stating, "Once luck has been stirred in the whole idea of progress driven by some innate striving, or superiority, or destiny, becomes nonsense,"[38] he has left the realm of biology and has emerged as an amateur philosopher. What scientific test has ever been done to show it is "nonsensical"? After all, there is reasonable evidence, even of the scientific sort,[39] to think that the bodily resurrection of Jesus really did take place. Since that is so, then surely it is also reasonable to think that we are here through divine providence. Gee's "nonsense hypothesis," Wilson's "luck hypothesis," and Gould's "tape-replay hypothesis" have not been tested in any meaningful way, and the scientific enterprise as a whole ought to remain silent or at least admit its philosophical biases until it is able to put these specific hypotheses about our origin to the test.

In complete contrast to the conclusions of Gee, Wilson, Gould, and many other biologists, scientific data have been emerging that are highly consonant and beautifully consistent with the Christ who is "before all things and through whom all things hold together" (Col. 1:17 New Revised Standard Version), and the Word "through whom all things came into being . . . and without him not one thing came into being" (John 1:4 New Revised Standard Version). Francis Collins writes in *The Language of God*:

> Altogether, there are fifteen physical constants whose values current theory is unable to predict. They are givens: they simply have the value that they have. This list includes the speed of light, the strength of the weak

38. Henry Gee, *The Accidental Species: Misunderstandings of Human Evolution* (Chicago: University of Chicago Press, 2013), 13.

39. Historical science of the sort outlined by N. T. Wright or Larry Hurtado is a legitimate way of knowing too. Indeed, evolutionary biology is in no small part a historical science. See N. T. Wright, *The Resurrection of the Son of God: Christian Origins and the Question of God* (Minneapolis: Fortress, 2003). See also Larry Hurtado, *Lord Jesus Christ: Devotion to Jesus in Earliest Christianity* (Grand Rapids: Eerdmans, 2003).

and strong nuclear forces, various parameters associated with electro-magnetism and the force of gravity. The chance that all of these constants would take on the values necessary to result in a stable universe capable of sustaining complex life forms is almost infinitesimal. And yet those are exactly the parameters that we observe. In sum, our universe is wildly improbable.[40]

Against all odds, it seems, the conditions in the universe that emerged from the big bang were exactly those required for life. And now, more recently, biology has shown that even with those parameters intact, the probability of our coming into existence as a species would appear to have been near zero. But the natural scientists are limited by the tools at their disposal for exploring the origin of humankind. Those tools bring us to a frontier that is wonderfully consonant with Christian theology and the Hebrew scriptures. The knowledge acquired from the scientific frontier opens up the opportunity for a whole new set of tools based upon a different supposition: there *is* such a thing as divine providence. Given this supposition, the tools to explore its ramifications are theological and philosophical in nature. The frontiers these tools open up are every bit as exciting and likely even more important as they address questions foundational for the very meaning of human existence.

40. Francis S. Collins, *The Language of God* (New York: Free Press, 2006), 74.

2 In Adam All Die?

Questions at the Boundary of Niche Construction, Community Evolution, and Original Sin

CELIA DEANE-DRUMMOND

The relationship between evolution and the Roman Catholic tradition can best be described as edgy. Karl Rahner notes that the early total rejection of evolution from the middle of the nineteenth century to the first decades of the twentieth in some cases went as far as to claim that evolution was religious heresy.[1] In Jack Mahoney's history of the reception of evolution in official Catholic teaching, he observes that the Pontifical Biblical Commission issued a statement in 1909 that insisted on the literal and historical truth of the first three chapters of Genesis.[2] By 1941, Pope Pius XII was still not convinced that evolutionary science had proved anything definitive about human origins, but by 1948 the Biblical Commission shifted to a somewhat more qualified position, resisting the idea that it would be misleading to reject history as relevant to the early chapters of Genesis on the basis of literary genre. Pope Pius XII's encyclical

---

1. Karl Rahner, *Hominization: The Evolutionary Origin of Man as a Theological Problem* (London: Burns & Oates, 1965), 29. See also Celia Deane-Drummond, *The Wisdom of the Liminal: Human Nature, Evolution and Other Animals* (Grand Rapids: Eerdmans, 2014).

2. I am indebted to Mahoney's survey for many aspects of the précis of the relationship between Catholicism and evolution. Jack Mahoney, *Christianity in Evolution: An Exploration* (Minneapolis: Fortress, 2011).

I would like to thank all my colleagues in the Colossian Forum colloquia for their extremely helpful feedback, along with the editors Jamie Smith and Bill Cavanaugh. I would also like to thank my research assistant, Craig Iffland, who made many perceptive editorial comments in the final stages.

Humani generis,[3] released a few years later in 1950, was a major statement on evolution, but he separated evolutionary theory, which he believed could apply to the human body, and religious teaching on the soul, so that "catholic faith bids us hold that souls are immediately created by God" (§36). But, significantly, he also rejected the idea that Adam could be polygenic, a community representative. That insistence seems to be related to his wish to preserve the church's traditional teaching on original sin.[4] It is worth citing *Humani generis* here, for I believe it is relevant to subsequent developments in Catholic reflection:

> For the faithful cannot embrace that opinion which maintains that either after Adam there existed on this earth true men who did not take their origin through natural generation from him as from the first parent of all, or that Adam represents a certain number of first parents. Now it is in no way apparent how such an opinion can be reconciled with that which the

3. Pope Pius XII, *Humani generis* (1950), http://www.vatican.va/holy_father/pius_xii/ encyclicals/documents/hf_p-xii_enc_12081950_humani-generis_en.html. He also describes the idea that the world is in continual evolution as a "pantheistic" notion, accusing such views as being "materialist" and resisting "all that is absolute, firm and immutable," opening the way for other "existentialist" dangers that resist belief in "immutable essences" (§5–6).

4. It is interesting to note that Karl Rahner also resists the idea of *polygenesis*, claiming that "polygenism as an object of divine action is impossible" (Karl Rahner, "Theological Reflections on Monogenism," in *Theological Investigations*, vol. 1: *God, Christ, Mary and Grace*, trans. Cornelius Ernst [London: Darton, Longman & Todd, 1961], 291). He is able to accommodate this view with his more open reception to evolutionary theory by labeling the early books of Genesis as prehistorical or primordial history and so outside what he considers are the proper boundaries of science, rather than historical. His argument against polygenism is supplemented by three theological arguments. First, he insists that the first man is the institution of a prime cause, not the institution of an effect (292). Second, that the becoming human is a miracle that must have taken place only once, rather than multiple times, because it established a new metaphysical condition, after which it would multiply by itself (295). Third, he argues that it cannot be proved that human beings would not have been able to continue if they had first appeared as a single pair (298). See Rahner, "Theological Reflections on Monogenism," 229–97. The third argument is now incorrect based on genetic evidence; see other chapters in this volume. His second argument assumes that God necessarily acts individually rather than through a group. However, this presupposes a post-Enlightenment view of the human as a single individual. There is no reason, it seems to me, why a new metaphysical condition could not have begun in a group. The first argument that this was an institution of a prime cause presupposes a necessary and miraculous divine intervention at the dawn of humanity that is metaphysically different from any other act of God's creation. This splits apart human becoming from the rest of evolution, and posits God as acting in the gaps of scientific knowledge; namely the mystery of the difference between humans and other animal kinds.

24

sources of revealed truth and the documents of the Teaching Authority of the Church propose with regard to original sin, which proceeds from a sin actually committed by an individual Adam and which, through generation, is passed on to all and is in everyone as his own.[5]

By 1966, Pope Paul VI was prepared to call evolution "a theory" but still largely repeated the message of *Humani generis*. In 1996, Pope John Paul II was ready to acknowledge evolution as *more than* a *mere* hypothesis, even while insisting on the mystery of what he termed the "ontological leap" between animals and human beings. Such a difference, he believes, refers to the spiritual realm which can be signified by the sciences, but the experience of which falls within the proper task of philosophy and ultimately theology.[6]

What is of special significance here is the tone of John Paul II's readiness to take the sciences seriously and acknowledge the different evolutionary theories contributing in a valid way to a discussion about the human person. Given that, it is hardly surprising that many theologians have experienced profound shock that the 1994 Catechism of the Catholic Church does not mention evolution in its discussion of creation and human origins.[7] Daryl Domning describes such an omission as scandalous.[8] Mahoney has gone beyond such shocked accounts to try and discover why the Catechism failed to mention evolution. He traces this omission to the influence of Cardinal Christoph Schönborn, who believed teaching on creation was too often passed over for fear of conflict with science, including the topic of original sin. In this matter, Schönborn insisted that the Catechism could not present "novel theses." But his lack of reference to evolution also needs to be considered in the light of his apparent adherence to intelligent design; a view

---

5. Pius XII, *Humani generis*, §37.

6. This was issued in a statement to the Pontifical Academy of Sciences. For a reproduction of that statement, as well John Paul II's "Letter to George Coyne, S.J." (Director of the Vatican Observatory) on his approach to theology and science, and my commentary on both statements, see David Marshall, ed., *Science and Religion: Christian and Muslim Perspectives* (Washington, DC: Georgetown University Press, 2012), 152–72.

7. Gabriel Daly describes such an omission as disregarding difficulties of modern believers, and comments on its description of the creation of human beings and fall from grace as an odd mixture of historical literalism and admission of symbolic language. Gabriel Daly, "Creation and Original Sin," in *Commentary on the Catechism of the Catholic Church*, ed. Michael Walsh (London: Geoffrey Chapman, 1994), 82–111, see especially 94–104.

8. Daryl Domning, "Evolution, Evil and Original Sin," *America* 185, no. 15 (November 2001): 14–21.

that is profoundly ironical in light of his refusal to allow Catholic theology to be swayed by new ideas.[9]

What this brief introductory summary shows is that while the door to dialogue with evolutionary theory opened up in official Catholic teaching, it has been influenced by a rather more conservative backlash. The stumbling block seems to be located squarely on human origins in general, with particular worries about reducing humanity to material nature, alongside some concerns about original sin.[10] Jack Mahoney has reacted to attempts to shore up a conservative interpretation of Catholic teaching on evolution by rejecting both original sin and the Fall.[11] He arrives at such a conclusion by way of a new Christology; Christ redeems the human race from death and self-centeredness that for him is the inevitable fate of human evolution. Jesus's death was therefore "a cosmic achievement for humanity, taking our species through the evolutionary cul-de-sac of individual extinction to enter into a new form of human living."[12] In rejecting original sin, Mahoney rejects both "the idea of an early collective lapse of the whole of humanity from divine grace through an original sin of its proto-parents," and "that God became man in order to restore through his death this fallen humanity to its original state of divine friendship."[13] He traces the Catechism's interpretation of original sin as primarily indebted to Augustine's faulty exegesis of Romans 5:12 that then allowed for an interpretation of original sin through propagation, and in Augustine's case, through sexual desire and generation rather than through imitation.

Mahoney is quite correct to take evolutionary theory seriously in considering human origins. However, he may not have taken it quite seriously enough. In the first place, while recognizing that within evolutionary theory there are also altruistic tendencies in other species that take a distinctive moral form in human beings, he seems to forget such an evolutionary trajectory when he discusses Christ's redeeming grace as a kind of cosmic reversal of the inevita-

---

9. While not all advocates of intelligent design oppose every element of evolutionary theory, at least on the surface, Schönborn clearly does hold such a view, dismissing John Paul II's letter to the Academy of Sciences as "rather vague and unimportant." See Christoph Schönborn, "Finding Design in Nature," *New York Times*, 7 July 2005.

10. Retaining a strong sense of human immateriality is likely to be even more vehemently defended in traditional Roman Catholic circles compared with specific interpretations of original sin, partly because of John Paul II's insistence on the "gap" between human beings and other animals, and also because without such immateriality original sin fades from view.

11. Mahoney, *Christianity in Evolution*, 51–57 and 72–74.

12. Mahoney, *Christianity in Evolution*, 51.

13. Mahoney, *Christianity in Evolution*, 52.

bly selfish banner of evolution. While the selfishness myth is strident in popular evolutionary theory, to contend that Christ is the answer for the human race to such a fate is just as difficult to believe from a scientific perspective as that Adam originated human sin. He is, however, quite correct in my view to resist a strictly Augustinian view of sin and guilt as propagated through sexual intercourse and direct generation, for if that were the case then Christ would have been infected through a sinful lineage, given the messiness in ancestral relationships that preceded his birth.[14] It is certainly difficult to prove Mahoney's contention that Augustine's focus on sexuality is related to his particularly troubled struggle with chastity and sexual desire. But where Mahoney goes particularly wrong, it seems to me, is not just in simplifying the evolutionary account but also in rejecting outright any belief that human sin had a beginning in history. The difficulty, of course, is what kind of history that entails. Mahoney is also erroneous in considering human evolution as if it were an isolated species.[15]

How far reactions to conservative hesitation about evolutionary biology will change under the papacy of Pope Francis will be interesting to behold, since it is known that he affirms the concept of evolution, rather than taking a purely literalistic view of the Genesis account. Pope Francis therefore claimed in an address to the Pontifical Academy of Sciences:

14. The Catholic tradition of the immaculate conception of Mary that taught she avoided the stain of original sin emerges, ironically perhaps, in 1854, around the same time as Charles Darwin's theory of evolution. It could be perceived as an attempt to get around difficulties in coming to terms with a biological notion of original sin and the human vessel through whom Christ became incarnate. But the question could be pushed back to absurdity. For example, "How was it that Mary's parents were able to conceive immaculately?" and "Why is it that Jesus's lineage in Jewish history is traced through Joseph, who is the adopted father of Jesus, rather than his biological father?" Such questions do not get very far in speaking about the mystery of the incarnation, and seem to stem from a mistaken view of original sin through propagation. Other traditions that put less weight on the importance of Mary's immaculate conception manage to avoid this problem. Aquinas, for example, believed that the manner of Christ's birth avoided original sin, even though he took on the "flesh" of Adam, as he assumed, with the tradition, that there was no semen involved in Christ's incarnation (see Aquinas, *Summa theologiae*, III q. 15.1 ad. 2 and q. 31.1 ad. 3), and so that incarnation was not in any way contingent on Mary's immaculate state. The clash between this position in relation to contemporary biology remains, but this time it is associated with what seems to be an a-biological origin of Christ's humanity.

15. In purely evolutionary terms human beings are actually not all that well adapted; molecular biology points to humans as an overspecialized species with many maladaptive features rather than as a pinnacle of evolutionary achievement. Mahoney tends to rely on the mid-twentieth-century version of evolution according to the Modern Synthesis, which places pan-adaptionism as its central vehicle for evolution.

When we read the account of Creation in Genesis we risk imagining that God was a magician, complete with an all-powerful magic wand. But that was not so. He created beings and he let them develop according to the internal laws with which He endowed each one, that they might develop, and reach their fullness. He gave autonomy to the beings of the universe at the same time in which He assured them of his continual presence, giving life to every reality. And thus Creation has been progressing for centuries and centuries, millennia and millennia, until becoming as we know it today, precisely because God is not a demiurge or a magician, but the Creator who gives life to all beings.[16]

Gone then is any possibility that intelligent design is a viable option. There is still a novelty and difference between human beings and the rest of the natural order, namely, the capacity for freedom, so later in the same address the pope states, "God gives the human being another autonomy, an autonomy different from that of nature, which is freedom." He avoids, nonetheless, the trickier question of original sin, but rather speaks of empowerment of humanity by God to make a better and livable world. Sin, in this view, seems to be a misplaced autonomy rather than freedom under God's direction. He also, significantly in my view, speaks of the need for a greater sense of human responsibility for stewardship of the natural world, showing a keen awareness of pressing matters related to environmental problems. *Laudato Si'*, a papal encyclical released by Pope Francis in 2015, reinforces such a claim and is broadly in line with Pope John Paul II's position, such that he holds to an evolutionary view of the biological world, while resisting any suggestion that this is sufficient to account for human uniqueness. Thus Pope Francis states,

> Human beings, even if we postulate a process of evolution, also possess a uniqueness which cannot be fully explained by the evolution of other open systems. . . . Our capacity to reason, to develop arguments, to be inventive, to interpret reality and to create art, along with other not yet discovered capacities, are signs of a uniqueness which transcends the spheres of physics and biology.[17]

16. Pope Francis, "Address of His Holiness Pope Francis on the Occasion of the Inauguration of the Bust in Honour of Pope Benedict XVI," 27 October 2014, https://w2.vatican.va/content/francesco/en/speeches/2014/october/documents/papa-francesco_20141027_plenaria-accademia-scienze.html.

17. Pope Francis, *Laudato Si': On Care for Our Common Home* (24 May 2015): §81,

Of course, he is not ruling out here the contribution of the social sciences, especially anthropology, which would similarly resist ontological reductionism. It is therefore extremely doubtful that Pope Francis will ever go as far as Mahoney and argue that evolution provides a sufficient explanation for the origins of human suffering and death, so there is now no need for the doctrines of the Fall and original sin. It is worth noting that where he does talk about sin in *Laudato Si'*, it is in the context of the breakdown of relationships with God, each other, and earth, thus including sins against the created order.[18] Such mythologies in Mahoney's view reflect the obsession with sinfulness in the Jewish tradition, born from centuries spent in exile and oppression.[19] But can this account of the origin of sin be dispensed with so readily on the basis of evolutionary theory or even biblical interpretation?

Revisiting the Fall

Attempts to debunk original sin raise wider questions about methodology: How far and to what extent should theologians seek to respond to scientific theories that may themselves become obsolete over time? It is not desirable, it seems to me, to use scientific theories as some sort of epistemological grid through which theological ideas have to pass in order to be acceptable. This concedes far too much to the dominating status of science in the academy. At the same time to ignore evolutionary science altogether in the interests of epistemic purity is naïve, especially in cases where there are converging interests and public debate on topics such as human evolution. The crucial question, of course, is what criteria help adjudicate which theological stances are to be retained and which rejected. Traditional Catholic thought has tended to rely on the Magisterium for this process, but, as I have outlined above, not all theologians are prepared to accept this authority, and where the Magisterium is ambiguous confusion abounds.

Ernan McMullin presses for classical criteria to be used in the interpretation of scripture in the light of science, the first of which is prudence.[20] His

---

http://w2.vatican.va/content/francesco/en/encyclicals/documents/papa-francesco
_20150524_enciclica-laudato-si.html.

18. Pope Francis, *Laudato Si'*, §2; §8; §66; §239. He also speaks of the "sin of indifference" in the prayer closing the encyclical.

19. Mahoney, *Christianity in Evolution*, 62–63.

20. Ernan McMullin, "Galileo on Science and Scripture," in *The Cambridge Companion to Galileo*, ed. Peter Machamer (Cambridge: Cambridge University Press, 1999), 292–99.

second criterion of priority of demonstration states that in cases of conflict between a "proven truth about nature" and an interpretation of scripture, scripture should be reinterpreted. Of course, it is somewhat problematic to use his guidelines that refer to "scripture" and make this apply to theological statements such as original sin. Even if we rely on this criterion for scripture, it would mean that belief in the resurrection would be difficult. Thirdly, and surprisingly perhaps for a Catholic theologian, he puts a high priority on scriptural interpretation, so that where a conflict exists with other teachings a literal interpretation from scripture should prevail. But there are thorny questions about precisely what this means, given the complex history of hermeneutics. His fourth criterion that the primary concern of scripture is salvation, not science, is a good qualifier, and sets the tone of any engagement with scientific work. My own methodological presupposition is that engaging with relevant sciences enriches theology and, more radically, vice versa. The extent of that enrichment will not necessarily be the same, and may be limited to pushing science to ask different kinds of questions, but the possibility of reciprocity is there nonetheless. McMullin does not consider this latter possibility here even though he is very attuned to historical examples that show theological beliefs were important for critical breakthroughs in science, Isaac Newton being a good example. Further, to limit theology to salvation puts far too much emphasis on salvation history at the expense of proper due consideration to the created world as such.

Hence, I suggest that a theological account can help open up new questions for evolutionary theory, especially in areas where that theory is still developing. I do not think theologians need to be shy about the fact that in due time science will change. Theology is to be written anew in every generation, even if that means that it is in need of constant revision. The judgment of which aspects to keep and which are adjustable is not just that according to prudence, or practical wisdom, which McMullin suggests, but also that according to the virtue of wisdom, which Aquinas believed was the intellectual virtue of speculative reason, and therefore particularly relevant for theology.[21] So knowledge of the highest cause, as relevant to theology, is related to the capability of wisdom as such and not simply prudence.[22]

Mahoney's rejection of the early chapters of Genesis in the name of sci-

---

21. Thomas Aquinas, *Summa Theologiae: Consequences of Charity*, vol. 35 (2a2ae 34–46), trans. Thomas R. Heath (London: Blackfriars, 1972), 2a2ae q. 45 a. 1.

22. Thomas puts it succinctly thus: "In man different objects of knowledge imply different kinds of knowledge: in knowing principles he is said to have 'understanding,' in knowing conclusions 'science,' in knowing highest cause 'wisdom,' in knowing human actions 'coun-

ence overlooks the tendency to conflate the philosophical distinction between the origins of mortality and finitude with the origins of sin. Edward Farley has dealt with the latter in his *Good and Evil: Interpreting a Human Condition*. Like Mahoney, he rejects a traditional account of original sin and the Fall, but still retains features of the classical account. The first, and most important, is what he terms one of the seminal insights of the Hebraic tradition: differentiating sin from the tragic; a point also noted by Paul Ricoeur when he distinguished the Adamic from the tragic myth.[23] Thus Farley points to a crucial issue in the classical account that is completely missing in Mahoney's account, namely, that "[b]ecause inter-human violations (evil) were not just fateful inevitabilities, they called not just for ritual protection but for resistance and change. And with this differentiation between the human tragic condition and human sin came a new sense of salvation and the notion of history itself."[24] He also points to the second feature of the classic view that relies on theo-centrism. This means that the interpretation of what is driving acts of human evil is a passion gone wrong, so "[s]in (moral corruption, oppression, inter-human violation) arises from a skewed passion for the eternal, in other words, idolatry."[25] Worship of what is truly eternal is the means to break the dynamics of evil. The third aspect of the classical view that Farley affirms, also correctly in my view, is that sin has *ontological* ramifications, and so is an alteration in the *being* of human agents that modifies the structures of the self. Hence, it is not simply about observable behaviors. By "being" he means "the dimensions of self-presencing, the biological, the passions, as well as by participation in the spheres of the interhuman and the social."[26]

There are two features of the classical view that Farley fiercely rejects: (a) the comprehensive cosmological narrative framework and (b) the quasi-biological explanation of sin's universality. The important point in (a) is that when sin is placed in this cosmic narrative it has a prehistorical origin through the rebellion and Fall of Satan, a historical origin, and the promise of future redemption. Karl Rahner largely accepts this cosmological narrative, and uses it in order to affirm the biblical accounts of the origin of human-

---

sel' or 'prudence.'" Thomas Aquinas, *Summa Theologiae: Knowledge in God*, vol. 4, trans. Thomas Gornall (Cambridge: Blackfriars, 1964), 1a q. 14 a. 1.

23. Paul Ricoeur, *The Symbolism of Evil*, trans. E. Buchanan (New York: Harper & Row, 1967).

24. Edward Farley, *Good and Evil: Interpreting a Human Condition* (Minneapolis: Fortress, 1990), 126.

25. Farley, *Good and Evil*, 126.

26. Farley, *Good and Evil*, 127.

kind, while also engaging with evolutionary science. So, for Rahner, the early origin story about humankind is *primordial* history that, for him "lies outside the scope of natural science," and "has a certain historical transcendence and cannot be examined as if it were one element among others in our history. Of their very nature, the reality of primordial history and eschatology is farthest removed from our idea of them."[27] But while Rahner could make such claims in the light of scientific ignorance about human origins, in the last quarter-century the field of evolutionary anthropology has grown exponentially. It is no longer sufficient to wriggle out of the engagement by labeling early human origins prehistory, even if the scientific basis for some of the claims is more speculative compared with the harder natural sciences. Rahner is, however, prepared to adjust the traditional neo-Augustinian Catholic view by recognizing (like Farley) that sin is not simply the introduction of mortality; rather, "original sin simply means that man, because he is a descendant of Adam, belonging to this historical, human family, ought to possess divine grace but does not do so."[28]

Farley's second area for concern is the neo-Augustinian belief that sin persists in history through a corrupted human nature that is passed down through biological propagation. But is Farley fully consistent in rejecting *any* notion of historical origin in relation to sin? For he concedes that the very idea of history and differentiation between the tragic and interhuman violations have an origin in time. Further, if evolutionary history is used as a more prominent backdrop to the stories of the origin of sin, it becomes clear that there is a gradual awakening to the religious in human prehistory millennia before it comes to be recorded in the book of Genesis. How might evolution and the Fall intersect in a richer way compared with that given in Farley's account? In particular, like Mahoney, he gives far too much attention to the human species and does not take the creaturely context of the human seriously enough.

Evolution and the Fall

The first important point to note about the evolutionary history of human origins is the need for such origins to be understood primarily in the context of

27. Karl Rahner, "Original Justice," in *Concise Theological Dictionary*, 2nd ed., ed. Karl Rahner and Herbert Vorgrimler (London: Burns & Oates, 1983), 354 (353–54).

28. In other words, for Rahner it is the absence of openness to grace that leads to original sin. See Karl Rahner, "The Body in the Order of Salvation," in *Theological Investigations*, vol. 17: *Jesus, Man and the Church*, trans. Margaret Kohl (London: Darton, Longman & Todd, 1981), 73.

a community.[29] The issue here relates to a relatively new emphasis on niche construction in theories of human evolution.[30] Standard neo-Darwinian theories that have been most influential in public reception of evolutionary theory envisage organisms as possessing traits that are most suited to their external environment. Niches or the wider communities in which organisms are located are of purely ecological rather than evolutionary interest in this view. The idea that organisms inhabit a wider world or *Umwelt* has been known for some time. So there is a focus in standard theories on the selection of given traits or characteristics. Once niche construction becomes part of the *evolutionary* narrative, organisms are viewed as actively influencing their environment rather than simply passive in response to it according to the filtering criteria of natural selection.

Jeremy Kendal et al. sum this up well when they suggest that "the defining characteristic of niche construction is not the modification of environments per se, but rather the organism-induced changes in selection pressures in environments."[31] Niche construction theory (NCT) represents an important philosophical shift in the way evolutionary processes work, so that evolutionary questions are viewed under the umbrella of NCT, rather than simply adding it on to previous models.[32] Standard evolution theory is "externalist" inasmuch as the environment is viewed as external factors that are acting in order to select those internal properties that are most adapted to that environment. Many theological commentators on evolution assume the standard *externalist* model of human evolution, where death is the decisive factor in natural selection, eliminating those individuals that are weakly adapted to the environment. Natural selection in this view is the "ultimate" category that

29. For further elaboration of hominin evolution see Celia Deane-Drummond, "Evolutionary Perspectives on Inter-Morality and Inter-Species Relationships Interrogated in the Light of the Rise and Fall of *Homo sapiens sapiens," Journal of Moral Theology* 3, no. 2 (2014): 72–92.

30. See Celia Deane-Drummond and Agustin Fuentes, "Human Being and Becoming: Situating Theological Anthropology in Interspecies Relationships in an Evolutionary Context," *Philosophy, Theology and the Sciences* 1 (2014): 251–75.

31. Jeremy Kendal, Jamshid J. Tehrani, and F. John Odling-Smee, "Human Niche Construction in Interdisciplinary Focus," *Phil. Trans. Royal Society B* 366, no. 1566 (2011): 785–92.

32. The mathematical expression of NCT is straightforward. Standard evolutionary theory assumes an organism's state is a function of the organism and the environment ($dO/dt=f(O,E)$) and changes in the environment are simply a function of that environment ($dE/dt=g(E)$). NCT, on the other hand, allows for the organism to be able to change the environment, and so can be expressed mathematically as $dO/dt=f(O,E)$ and $dE/dt=g(O,E)$.

explains phenotype, including behavioral differences, and therefore devalues more "proximate" causes. Standard evolution theory can still include NCT, but the "ultimate" explanation is still one rooted in natural selection.

In the newer approach to NCT, the idea of some sort of direct evolutionary "causation" through natural selection becomes problematized. So the "dichotomous proximate and ultimate distinction" is replaced by "reciprocal causation."[33] In this way, niche construction works *with* natural selection in the evolutionary process in a dynamic interchange. Niches are themselves part of the inheritance process, so that an *interactionist* evolutionary theory replaces an *externalist* theory. Niche construction emphasizes not just genetic and cultural inheritance, but also ecological inheritance as well, in dynamic interaction with the first two.[34] But even in this model, envisioning cultural aspects as somehow separate from ecological inheritance seems too constraining. Ecological and cultural inheritance under a broader "ecological" category carries the advantage of perceiving a developmental context in which the physical niche is not separated from the social niche.[35]

This dynamic interactionist view is relevant for a theological discussion of the origin of sin. It removes, in the first place, the determinism often characteristic of externalist evolutionary accounts. Instead, creatures become active agents of their own evolution, even if such agency is only capable of being perceived to be such retrospectively by a tiny fraction of creaturely beings, namely *Homo sapiens sapiens*. Agustin Fuentes envisages that in the course of hominin history the *Homo* community niche gradually shifted.[36] He suggests that rather than simply using anatomical markers, the function of communities in the evolution of the genus *Homo* allow for four different community niches to be distinguished, namely (1) early *Homo* (2.3 mya–1.5 mya), (2) *Homo erectus* (1.5 mya–0.7 mya), (3) erectus/archaic *Homo* (0.7 mya–0.3 mya), (4) modern humans. In this model the earliest communities that were characteristic of *Homo* species existing over 1.5 million years ago were small and static communities and suffered numerous extinctions. Over time these moved eventually to larger, more communicative, interac-

33. Kendal et al., "Human Niche Construction," 786.

34. Kevin N. Laland, F. John Odling-Smee, and Marc W. Feldman, "Cultural Niche Construction and Human Evolution," *Behavioral Brain Sciences* 23 (2000): 131–75.

35. F. John Odling Smee, "Niche Inheritance," in *Evolution: The Extended Synthesis*, ed. Massimo Pigliucci and Gerd B. Muller (Cambridge, MA: MIT, 2010), 175–207.

36. Agustin Fuentes, "Human evolution, niche complexity, and the emergence of a distinctively human imagination," *Time and Mind* 7, no. 3 (2014): 241–57. See also Deane-Drummond and Fuentes, "Human Being and Becoming."

tive, and far more flexible groups. Symbolic and cultural innovation in this model appeared prior to the development of full-blown symbolic language in the transition between niche 3 and niche 4. Fully organized religion did not appear until much later, 10,000 years ago, so could not account for this transition. The capability for imaginative experience, including latent forms of religious experience, appeared very much earlier than the first recorded evidence of either early cave paintings or organized religious beliefs. The standard view that the most significant shift from anatomically modern humans to behaviorally modern humans was constricted to *Homo sapiens* is beginning to come under scrutiny because of archeological evidence of fainter traces of such activities in other hominins. It therefore seems to be mistaken to presume that there was a single transition in behaviorally modern humans, but that, fascinatingly, glimpses of what seem eventually to be characteristic of our own species appeared much earlier than that in those hominins pre-dating *Homo sapiens*, such as *Homo ergaster* and *Homo erectus*.[37]

If the Fall is understood as a simple spreading of destructive behavior, then there is ample evidence to suggest that early human communities learned by imitation of others in a way that was passed down from one community to the next, and thus need not depend on language as such. It seems to mean more than this, namely, a deliberate turning away from God and a prideful assertion of self as God, *sicut deus*, in its place, what Farley names as idolatry. Some evolutionary biologists have even tried to account for the emergence of religion through what might be called an "original sin" model, where those who fail to cooperate in a community ("sinners") are in need of punishment, and therefore God is invented by that community in order to provide a convenient and very cost-effective way of dealing with that difficulty.[38] In this case, the temptation to do wrong against others in a community comes *first*, what is known in evolutionary game theory parlance as "free riders," and then religious belief follows in order to suppress that tendency. The traditional Genesis account of the Fall of the first human pair inverts this view entirely: the relationship with God is primary and right relationship with God is realized in the Garden of Eden. The mistake of evolutionary biologists who add God onto their account of "free rider" explanations is that in their deliberations God is assumed to be perceived as

37. Marc Kissel and Agustin Fuentes, "From Hominid to Human: The Role of Human Wisdom and Distinctiveness in the Evolution of Modern Humans," *Philosophy, Theology and the Sciences* 3, no. 2 (2016): 217–44.

38. Dominic Johnson, "Why God Is the Best Punisher," *Religion, Brain and Behavior* 1, no. 1 (2011): 77–84.

being just like other agents. This does not make sense theologically, since God is the ground of being, rather than a Being alongside other beings. The conscious communication between human beings and God at the dawn of human religious experience must have preceded the possibility of deliberate sin against God.

Classic Catholic theological accounts of the Fall of humanity that have come to dominate popular discussion require some radical revision if they are to be tenable, in that they tend to assume some sort of biological inheritance of Adam's guilt, which Augustine believed was transmitted through the male seed.[39] Ironically, perhaps, biology is taken far too seriously in this account of original sin. Possibly one instructive use of biology in a contemporary interpretation is to draw on psychology in the way that Kierkegaard attempted, so that at the root of original sin there is an existential anxiety that then becomes part of every subsequent free decision.[40] Karl Rahner, on the other hand, even while rejecting polygenism, favors a socially mediated interpretation and argues that original sin refers to that situation of guilt *into* which every human decision is then made.[41] I suggest it is important to distinguish between original sin understood as a contextual *ontological condition* in which human beings find themselves, as Farley points out, and *moral sin* as those acts for which every human being is held to account. The latter takes place in the context of the former and makes it humanly impossible not to sin, but when viewed in this way avoids either the inappropriate imputation of guilt on neonates in the way Augustine assumed or an adherence to a historically literal Garden of Eden, Adam and Eve. My own view is that the significance of Genesis is "historical," but without implying literal figures of Adam and Eve or a literal paradisiacal state before the Fall. Rather, Adam, as from the earth, and Eve, the source of life, *stand for* the community beginnings of the human race and what it might have become, while taking "immortality" as it existed in the Eden account as figurative. In this sense,

---

39. Augustine, *The City of God*, Book XIV, trans. Marcus Dods, accessed online at http://www.newadvent.org/fathers/120114.htm. Elaine Pagels is one of many scholars castigating such a view of the origin of sin as "absurd." See Elaine Pagels, *Adam, Eve and the Serpent* (London: Weidenfeld & Nicolson, 1988), 109.

40. Søren Kierkegaard, *The Concept of Anxiety: A Simple Psychologically Orientating Deliberation on the Dogmatic Issue of Hereditary Sin*, ed. Reidar Thomte and Albert B. Anderson (Princeton: Princeton University Press, 1980).

41. Karl Rahner, "Original Sin," in *Sacramentum Mundi: An Encyclopedia of Theology*, vol. 4, ed. Karl Rahner, Cornelius Ernst, and Kevin Smyth (London: Burns & Oates, 1969), 328–34.

evolutionary history is still relevant, but without needing to posit a literal pair or dyad through which evolution took its course.

Reimagining the Fall in Light of NCT

How might original sin be interpreted in the light of NCT? In the reception of classic interpretations of the Genesis text Adam and Eve are more often than not viewed as an isolated pair that has resonated in literary and cultural history.[42] While later Christian communities understood this evil as taking the particular form of Satan, evil's original appearance as the snake tends to be forgotten. Biblical exegesis of the Genesis text is worth considering in more detail in this respect, since the snake in Genesis 3:1 is named as one of the wild animals God has made.[43] The snake is one of the creatures that humans have interacted with, but inasmuch as they have named this creature, they have exerted dominion over it. But it is important to note in this context that the snake as such is not intrinsically evil according to the biblical account, but rather *'ārûm*, "cunning," which is translated variously as crafty, shrewd, subtle, and intelligent. The snake is still one of God's creatures. Further, the word to describe this form of intelligence is also sometimes associated with wisdom, but it is a wisdom that can be directed to good or evil ends.[44] Aquinas terms such a distorted form of practical wisdom "sham prudence" when exercised by humans.[45] Prior to the discussion of the Fall of humanity, the snake appears as an ambiguous figure who might use craftiness for good or evil ends. From a literary perspective there is also a play on the word, so that there is an association between the cunning of the snake and human nakedness, *'ărûmîm*, mentioned a few verses earlier.[46] There is an important etymological difference between the way the pun works here and that between Adam and *'ădāmâ* (ground) that Middleton highlights. Whereas in the case of Adam there is ontological resonance, in the case of the snake's craftiness/naked homonym comparison the reader is

42. For an interpretive survey of such literature see Brian Murdoch, *Adam's Grace: Fall and Redemption in Medieval Literature* (Cambridge: D. S. Brewer, 2000).

43. See Richard Middleton, "Reading Genesis 3 Attentive to Human Evolution," in this volume.

44. Middleton, "Reading Genesis 3."

45. Thomas Aquinas, *Summa Theologiae: Prudence*, vol. 36, trans. Thomas Gilby (Oxford: Blackfriars, 1974), 2a2ae q. 55 a. 8.

46. Middleton, "Reading Genesis 3."

likely to be in shock. As Middleton points out, "a smart or prudent person would *never* go around naked or vulnerable."[47] The snake, inasmuch as it practices deception, is the opposite of vulnerable or innocent. The difficulty that Middleton raises, namely, how can humans be responsible and *yet sin*, a problem that also exercised Paul Ricoeur, is at least made more coherent by thinking of human beings as being in closely interwoven relationships with other animals, including snakes.

Middleton concludes that the snake represents that aspect of creation that mediates ethical choice. The question remains as to how far such representation is figurative or whether specific animals can also be considered to be more concretely involved. Niche construction theories of evolution are inclusive of the importance of other species, and this argument can be extended in order to make a further claim, namely, that they were actually instrumental in the emergence of human im/morality.[48] The significance of human interaction with the snake at the dawn of moral awareness is barely commented upon in the literature.[49] Yet, this is precisely what one would expect if human becoming is densely interlaced with the lives of other creatures.

The early church fathers were more acutely aware of the relational aspects of human communities compared with most contemporary theological accounts of the origin of sin. Maximus the Confessor, for example, acknowledged the crucial importance of the breakdown in communal relationships; so that the Fall was also associated with individualization and fragmentation.[50] But what is interesting in this context is that even Maximus considered that such individualization represented a return to a bestial state, so that "we rend each other like the wild beasts."[51] So, the unity that represented the ideal state among humans was the province of humanity alone. In this view, redemption represents the recovery of a lost unity. Augustine also acknowledged the impact of sin as a breaking up of community relations, an

47. Middleton, "Reading Genesis 3."

48. I have argued for the role of interspecies relationships in the evolution of morality in Celia Deane-Drummond, "Deep History, Amnesia and Animal Ethics: A Case for Inter-Morality," *Perspectives on Science and Christian Faith* 67, no. 4 (2015): 1–9.

49. For further discussion on the emergence of morality in the human community see Celia Deane-Drummond, "Evolutionary Perspectives."

50. Henri de Lubac, *Catholicism: Christ and the Common Destiny of Man* (San Francisco: Ignatius, 1988), 33.

51. Maximus the Confessor, *Quaestiones ad Thalassium*, quoted in de Lubac, *Catholicism*, 43.

aspect that seems to have been quietly forgotten in contemporary discussions of his work through an almost obsessive focus on his negative association of sin and sexuality. So, Augustine claims, "Adam himself is therefore now spread out over the whole face of the earth. Originally one, he has fallen, and breaking up as it were, he has filled the whole earth with the pieces. But the Divine Mercy gathered up the fragments from every side, forged them in the fire of love, and welded into one what had been broken."[52]

Thomas Aquinas's account of sin is also interesting in that he stresses the communal aspect of Augustinian thought that has been woefully neglected (even though it was more common in the early church), namely, the concept of a fractured human nature *as such* that is in need of Christ's redemption. Aquinas acknowledges the worth of the Augustinian position, namely, that sin is transmitted by physical descent, denying the Pelagian rejection of this view on the basis that this would undercut Romans 5:18.[53] Yet his proof of such transmission is interesting in that he considers it necessary to consider human beings both as individuals and as in community. Comparing the human community to the unity found in the church, he suggests that "we should consider the whole population of human beings receiving their nature from our first parent as one community, or rather as the one body of one human being. And regarding this population, we can indeed consider each human being, even Adam himself, either as an individual person or as a member of the population that originates by physical descent from one human being."[54] While Aquinas seemed to believe in a historical Adam and Eve, his position still makes sense even without such a literal reading of Genesis.

And rather than stress negative aspects of original sin, Aquinas speaks in terms of an *original justice* bestowed on the first human through a supernatural gift of God, one that was not just for Adam, but that was intended as a source of the whole of the human race. So, on this basis, through the free choice of the first human who sinned the gift of justice was lost, passing this lack onto future generations, so "the privation is transmitted to them in the way in which human nature is transmitted."[55] Aquinas agreed with Augustine that the transmission of original sin was biologically mediated, but it was on the basis that human nature has a soul united to the flesh, and thus through

52. Augustine of Hippo, *On Psalm 195*, n. 15 (Patrologia Latina 37:1236), cited in full in de Lubac, *Catholicism*, 376.

53. Thomas Aquinas, *De Malo*, trans. Richard Regan (Oxford: Oxford University Press, 2001), q. 4 a. 1.

54. Aquinas, *De Malo*, q. 4 a. 1.

55. Aquinas, *De Malo*, q. 4 a. 1.

that flesh the gift of God is potentially propagated just as much as original sin. Any association of evil or indeed guilt with sexuality in the manner that Augustine presupposes does not make sense in Aquinas's account.[56] Further, he resists any sense that those who are born into original sin bear the guilt of Adam. Rather, it is only by considering the whole of the human race *as if it were one body* in association with the original voluntary moral fault of Adam that the privation of original justice becomes a moral fault. Sin consists in a disorder of a human will turned away from God and directed to a transient good, that itself bears an imprint and likeness to the first original sin.[57]

He also resisted the idea that other animals could have any kind of moral fault, and so they escaped the burden of original sin and the potentiality for original justice. Hence, "original sin inheres in the rational soul as the sin's subject."[58] And just as he believed that it was through the powers of the soul in *willing* that led to original sin, so something of the soul's essence becomes transmitted to future generations.[59] However, he did allow for a sharing in voluntary activity that is, at least, a component of the will. Like Augustine, he admitted to an outcome of original sin being an inordinate desire, so that a lack of original justice amounts to the quasi-formal element, while concupiscence corresponds with the quasi-material element. Both elements are in actual concrete sins as well in a way that mirrors original sin.[60] As I will suggest below, I believe we can nudge and enlarge his argument a little further in this respect in the light of what is now known about the close entanglement between human and other animals in community, and cognitive capacities in other animals.

Aquinas could be seen as resisting a wider medieval trend toward treating animals as if they were morally culpable in the same sense as human beings. David Clough, inspired by medieval historical court proceedings against other animals, argues that sin should be extended to include other creatures as well as human beings.[61] While I am sympathetic to the idea of

---

56. He claims, for example, in his replies to objections that "the reproductive act properly serves nature, since the act is ordained for the preservation of the species, and it belongs to the constitution of a human person that the flesh is already united to the soul." Aquinas, *De Malo*, q. 4 a. 1.

57. He discusses this aspect in more detail in the second article. See Aquinas, *De Malo*, q. 4 a. 2.

58. Aquinas, *De Malo*, q. 4 a. 3.

59. Aquinas, *De Malo*, q. 4 a. 4.

60. Aquinas, *De Malo*, q. 4 a. 2.

61. David Clough, *On Animals*, vol. 1 of his *Systematic Theology* (London: T&T Clark,

other animals having a moral agency of a kind, the definition of sin will impact on the extent to which other animals can share in sinful behaviors. Other animals can contribute to violent disruptions in community life, though that disruption is not deliberative nor does it have conscious reference to the divine in the manner that it can become in the human community. Instead, the proper relationship between human beings and other creatures is found in the way humanity is invited to name other animals prior to the Fall. Such a naming speaks of an ability to associate with other animals and create a world in which they have a significant place, rather than an occasion for human oppression.[62] However, if the story of the origin of humanity and sin in Genesis 1–2 is taken as a literal historical portrait it becomes difficult to square with modern evolutionary biology. Of course, many theological claims will not be fully compatible with evolution, and the tension remains, but it takes theological discernment to judge what should be insisted upon for theological reasons and what can be adjusted in the interests of scientific coherence.

Secular anthropology has also moved away from perceiving early hominins in terms of the isolated pair bond, to a much greater sense of the collective that is more characteristic of the earliest Christian traditions. Early *Homo* groups certainly seemed to be highly cooperative, and mass genocide was relatively rare, appearing later in the history of our species as the size of community groups exploded. However, it is perhaps forcing the account to associate the early pre-violent condition of *Homo sapiens* with any "paradisiacal" state.

What is interesting, nonetheless, is that the earliest account in the Genesis text associates paradise with harmonious relationships with other co-species and other creaturely kinds in a way that at least coheres with the perception of the lifeworlds of the earliest modern *Homo sapiens* species as entangled with each other and other species and in small supportive cooperative communities. Further, inasmuch as verbal communication apparently took place between God and these early human beings, some language must

---

2012), 104–21. I am rather less convinced than Clough seems to be that trials of animals tell us much more than the fact that anthropogenic habits of mind persisted in this period, that is, an inappropriate attribution of wrongdoing in other animals.

62. The fact that naming has, historically, been oppressive should not lead to the assumption that naming is inevitably like this. For further discussion, see David Clough, "Putting Animals in Their Place: On the Theological Classification of Animals," in *Animals as Religious Subjects: Transdisciplinary Perspectives*, ed. Celia Deane-Drummond, Rebecca Artinian-Kaiser, and David Clough (London: T&T Clark/Bloomsbury, 2013), 227–42.

have been possible, which confines the account to community niche 4, and thus to *Homo sapiens*, and most likely to *Homo sapiens sapiens*. It is therefore possible to align aspects of traditional interpretations of the biblical text with contemporary evolutionary theory, even while recognizing that all evolutionary and biological accounts have their limitations inasmuch as they are based on particular and perhaps peculiarly modern interpretations of what human beings are like. The objection that the Genesis text is not intended to be either a historical or a biological account of origins holds to the extent that the language used in Genesis is mythological rather than literal, but I suggest that an interpretation of such texts against a backdrop of contemporary knowledge of human becoming is nonetheless still illuminating. There is no need either to reduce Genesis to a crypto-science or to excise Genesis from consideration of the origins of sin and reduce it to contemporary evolutionary theory.

Recognition of that religious dimension in all things and particularly the capability for divine awareness as such would have dawned very slowly in human beings, but when it did, it brought with it the temptation for narcissistic self-reflection and pride in that difference, rather than using it to build relationships with God and each other. My own view is therefore that the Fall and original sin in particular show up the first occasion when divine awareness permitted an alternative self-destructive mode for human beings, leading to a recognized building up of hostility both within that community, and beyond that toward other animals as well. Human beings are in biological terms the most destructive animal on planet earth, for all their extreme abilities to do good and sacrifice for the sake of others. It is significant, in my view, that the charge to bear the image of God is not related to knowledge of good and evil, but to having correct dominion over the earth, a dominion that humanity has failed to exercise. Eden in this model represents the possibility that human beings were capable of divine recognition, but had no self-conscious evil tendencies. Before that there would have been some acts that harmed other persons, but they were more akin to that most commonly found in other animals, and therefore reactive and transactional rather than deliberate and transcendent. What is particularly interesting is that in contemporary tribal societies such as the Runa people, not only are other animals with whom they live in close association in the forest treated as subjects, but they have developed a transcendental explanation of the significance of these animals, especially the puma.[63] Such explanations seem

63. Eduardo Kohn, *How Forests Think* (Oakland: University of California Press, 2013).

to be partly related to the fact that these communities have to kill animals that they believe share in a subjective world. Such early hominins before the Fall would have built transactional social relationships, rather than the more imaginative and deliberative transcendental approach to social life.[64]

The thesis that I am proposing here, namely, that the ideal state should be viewed in community relationships, including multispecies relationships with other creatures, and that the Fall results in a distortion in those relationships, fits in with a strong sense of the moral collective that is common in smaller hunter-gatherer societies.[65] While awareness of individual sin would have been present, initially it was always filtered through a strong community sense. The notion that original sin is primarily about a distortion in community relationships is of course not new, and as I noted above appears in the early history of the church. De Lubac, therefore, interprets Maximus the Confessor as arguing for "original sin as a separation, an individualization it might be called in the depreciatory sense of the word."[66] The unity, therefore, that God intended is shattered by human sin. But Maximus's shattering of the possibility of human unity equates with becoming "like wild beasts." Given what we know about the sociality of other animals, such an assertion needs more qualification. What he can be taken to mean is that humans were no longer in control of their actions, their self-reflective powers diminished so as to represent types of actions found in other animals. Does that mean that other animals had no violence? Certainly not, for I am interpreting original sin in terms of self-conscious violence that is associated with the possibility of higher degrees of perception and religious awareness, rather than just raw violence against the community. It is also possible to interpret at least *some* violent tendencies in other animals as providing prerequisites toward what becomes the much more deliberative self-grasping character-

64. Maurice Bloch, "Why religion is nothing special but is central," *Philosophical Transactions: Biological Sciences* 363, no. 1499 (2008): 2055–61. Bloch is mistaken in reducing and diminishing religion to that which emerges simply from such transcendental capacities as if there were no possibility of revelation, for it fails to take sufficient account of religious experience, even if it is understandable that he holds this view from a secular perspective.

65. There is insufficient space to discuss this in any detail here. See Celia Deane-Drummond, "A Case for Collective Conscience: Climategate, COP-15, and Climate Justice," *Studies in Christian Ethics* 24, no. 1 (2011): 1–18.

66. De Lubac, *Catholicism*, 33. De Lubac references Maximus the Confessor, *Quaestiones ad Thalassium*, q. 2 (PG 90, 270); *De carit.*, century 1, n.71 (976) and century 2, n.30 (993) in making these remarks and the following comment, that "[w]hereas God is working continually in the world to the effect that all should come together in unity," but sin shattered this one nature.

istic of human beings.[67] So in this loose sense the shadow of the Fall is cast back in evolutionary history as well as forward into the future of human culture and subsequent development of sophisticated civilizations.

This shadow in other animal life is anticipatory of what happens in the Fall, and therefore the Fall should not be viewed as providing an explanatory account of "natural evil" as if there were no such evil before a human Fall, a classical view that Michael Murray has rightly rejected as no longer coherent.[68] The rise and fall of humanity should not be seen in detachment from other creatures, but in association and even in entanglement with them. The presence of natural evil is concomitant with the natural world as such, and navigating this problem raises much bigger questions about theodicy. However, distinguishing creation from nature through secondary causation provides some relief, for it is then no longer necessary to blame either God or humanity for everything that seems to be disrupted in the natural world.

Finally, it is worth emphasizing, with Orthodox theologian Kallistos Ware, who follows the early church fathers such as Athanasius of Alexandria (d. 373), that the fall of humanity should above all be thought of as a *process* of sinful degeneration, so that humans are born into a context where it is hard to do good and avoid evil.[69] While we might balk at the early church's

---

67. I therefore am prepared to argue that morality of a kind is found in other animals, including the possibility of ill. I prefer not to use the term "sin" for other animals, as this speaks of a more deliberative characteristic in relation to the divine. It would also be wrong to view what happens in other animals existing today as necessarily a simple precursor of what is found in humans, given the time since chimpanzees and other primates diverged from humans. I am prepared to stress the possibility of what I term intermorality, the evolution of human morality that is shaped in a significant way by the multispecies commons. For further discussion of animal morality, vice, and virtue, see Celia Deane-Drummond, *The Wisdom of the Liminal*, 122–52, and Celia Deane-Drummond, Agustin Fuentes, and Neil Arner, "Three Perspectives on the Evolution of Morality," in *Philosophy, Theology and the Sciences* 3, no. 2 (2016): 115–51.

68. Michael Murray, *Nature Red in Tooth and Claw* (Oxford: Clarendon, 2008), 73–106. Murray also includes a discussion of what he believes are the more convincing arguments that draw on pre-cosmic Fall narratives about Satan in relation to animal pain and suffering in order to provide an argument in defense of the goodness of God. While contemporary culture and more conservative Christians can adopt Satanic realism as a means to explain evil in the world, it becomes much harder for theologians to accept in the light of biblical exegesis of Genesis and post-Enlightenment thought.

69. Kallistos Ware, "The Understanding of Salvation in the Orthodox Tradition," in *For Us and Our Salvation: Seven Perspectives on Christian Soteriology* (IIMO Research Publication 40), ed. Rienk Lanooy and Walter J. Hollenweger (Leiden: Interuniversitair Instituut Voor Missiologie en Oecumenica, 1994), 113.

supposition that the fall is also a fall in physical capacities and death as well as moral capacities, the idea that human beings inherit tendencies toward wrongdoing through the damaged social context in which they are born is hard to refute. The early church was certainly not unanimous that this entailed, with Augustine, the inheritance of *guilt* as well as moral weakness. Ware concludes that while this theme is less well developed in the Eastern church, it is still present, that is, a recognition that insofar as we share in a collective sense of humanity, each and every one of us is called to repent of Adam's sin and that of our neighbor. But holding to a particular view of original sin is what could be termed a theologoumena, that is, not something that is required or necessary for Christian faith. My own view is that original sin can be reinterpreted to mean that a person is born in each generation into an imperfect community of others, including other creaturely kinds. That community shapes the particular tapestry of sin as expressed in the life of an individual sinner, where sin represents a cutting off from relations with God and with each other, leading to concrete wrongful acts to which each person can be held to account. It is not so much that guilt is inherited through original sin, but that original sin creates the distorted social context in which it is impossible not to be a sinner. Inasmuch as we are aware of a deep connectedness with one another's guilt, including the guilt of Adam (understood in a collective sense), there is also the possibility of original sin.

So in Christ Shall All Be Made Alive

A sense of the collective as well as individual nature of sin also impinges on how to consider redemption, and, in particular, the difficult passages in 1 Corinthians 15:22 and Romans 5:12 where Paul compares the death in Adam with the new life in Christ. The early church fathers viewed redemption as a bringing back of a unity lost through sin. So, in spite of Augustine's well-known discussion of the inner psychological distortion of the individual sinner, he also viewed sin and redemption through a collective lens. So "Divine Mercy gathered up the fragments from every side, forged them in the fire of love, and welded into one what had been broken. . . . He who remade was himself the Maker, and he who refashioned was himself the Fashioner."[70] This theme of repairing and healing what was ripped apart seems to be a constant refrain in the work of the early church fathers, so *Divisa uniuntur,*

70. Augustine, *On Psalm 195,* n. 15 (PL 37:1236), cited in de Lubac, *Catholicism,* 376.

discordantia pacantur.[71] There is, nonetheless, a problematic aspect of these classic accounts, and that is the assumption that suffering and the tragic follow from the fall into sin, rather than precede it. Evolutionary theory demands, then, an adjustment in such a perspective, so that sin takes place in a tragic context and the unity hoped for is an eschatological expectation of the end, rather than a return to a paradisiacal state. There are also dangers latent in the theme of unity if that unity becomes a fundamentalist form of oppression or a way of excluding diversity. That is why the Adam/Christ comparative passages in 1 Corinthians and Romans 5 are so significant in this context, for the unity experienced is one that allows flourishing of *all kinds* in Christ, including, it seems to me, all those creaturely kinds that are companion animals with human beings. The early church fathers generally held to a view of the human as ontologically and hierarchically distinct from other animals. Sin in this model is a sinking into the animal nature from an elevated position of suprarationality. There were some writers who were prepared to resist this model, notably, Irenaeus, Athanasius, and Basil of Caesarea, prominent theologians in the early church who pressed for an equivalent ontological status to be given to humans and other animals.[72] It is, in other words, a *liberative* rather than an oppressive unity. It is also a unity that speaks of the grace of God empowering what might seem to be impossibly peaceful relationships. So, as in Isaiah 11, the lion will lie down with the lamb and so on. This, too, like the Adamic passage in Genesis, is poetic, figurative anticipatory speech for peaceful coexistence among humans and all creatures, rather than suggesting a denial of basic teleology for different creatures. The lion will still be the lion, but the aggression that hampered life on earth will no longer be present in the future kingdom.

Now, it is quite possible, as many conservative exegetes have claimed, that in Romans 5:12 Paul did indeed think of Adam as a single individual in whose sin all humanity in subsequent generations participated. Rahner, as I indicated earlier, adopts this view for philosophical reasons, but he softens his position by naming Adam's origin as primordial history outside the range of science. The presupposition for this grace-filled condition in Adam is also

---

71. What was divided becomes united, discord becomes peace. De Lubac notes a number of different fathers who share this perspective. See de Lubac, *Catholicism*, 37.

72. Recognition of Basil of Caesarea's insistence on the ontological status of other animals, as well as the importance of creatures in the tradition of the early church has been surveyed in David Clough, *On Animals*, 26–27 and 45–49. Clough views the attempt to elevate human beings above other animals as a retrograde step in the tradition, as shown in Augustine, Aquinas, Maximus the Confessor, and many other early writers in the church.

redemption by Jesus Christ; in other words, for Rahner the plan for Adam was sanctifying grace that was only fulfilled through "bodily community of shared descent" in the light of Christ's coming.[73] Rahner seems to want it both ways here: historical and metaphysical, an insistence on one man, Adam, as a mirror of the grace in Jesus Christ.

But the important point here is the relationship between the individual and the collective, so that even in Rahner's position Adam is figuratively *representative* of the human race. But the analogy with Christ breaks down inasmuch as the salvation wrought in Christ through his divinity is effective for all humanity in a way that is not possible for a single prehistorical human being. Rahner tries to make Adam as significant as Christ by naming Adam in terms of a prime cause.[74] It seems to me that Rahner's notion of Adam's function as prime cause does not really make sense. This is because the disruption to the human race wrought by Adam's sin in the breaking of unity is not precisely the same as the healing of humanity and the reestablishment of unity through the work of Christ, otherwise Adam would have the quality of divinity. A prehistorical figure cannot be compared easily with the concrete historical figure of Christ, and inasmuch as the effects of Christ are transhistorical, they are transhistorical in a different manner compared with Adam's prehistory. Christ, through the work of grace, elevates humanity to the very life of God, whereas Adam disrupts the possibility of that relationship. While it is true that in Rahner's generation there was less known about early human origins, it is not really true in the present context. Adam marks what could be termed an ontological shift in the history of the human race, understood to have occurred at some period in prehistory, but most likely at a time when the human population was small. Dialogue with evolutionary theory opens up once again a thread recognized by the early church fathers, namely, the significance of human relationships with each other and the human set in the context of other creaturely kinds.

73. Rahner, "Body in the Order of Salvation," 73.
74. See more extensive discussion under footnote 4 above.

3 What Stands on the Fall?

A Philosophical Exploration

JAMES K. A. SMITH

In its (relatively) long encounter with evolutionary science, Christian theology has demonstrated a remarkable ability to absorb and accommodate new scientific consensus about cosmology, geology, and even human origins.[1] However, the traditional or orthodox[2] doctrine of the Fall has proven more difficult to reconcile with the picture of human origins that emerges with

---

1. Even "Old Earth Creationism" was able to absorb findings about the age of the earth. See Davis A. Young and Ralph F. Stearley, *The Bible, Rocks, and Time: Geological Evidence for the Age of the Earth* (Downers Grove, IL: InterVarsity, 2008) and, to some extent, pre-human evolution. See Mark Noll, *The Princeton Theology, 1812–1921: Scripture, Science, and Theological Method from Archibald Alexander to Benjamin Breckenridge Warfield* (Grand Rapids: Baker, 1983) and *Jesus Christ and the Life of the Mind* (Grand Rapids: Eerdmans, 2011), 99–124.

2. Granted, what I will describe as the "orthodox" understanding of the Fall is a "Western" (Augustinian) doctrine of the Fall but still, I would contend, a "catholic" inheritance. What I am calling the "traditional" or "orthodox" understanding of the Fall is a theological consensus across historic Protestant and Roman Catholic confessions (see, e.g., *Catechism of the Catholic Church*, 2nd ed., Part One, Section Two, Chap. 1, Para. 7, §§385–421). As such, the "orthodox" understanding of the Fall is not and should not be narrowly tied to modern "literalist" readings of Genesis characteristic of Protestant fundamentalism. Furthermore, I think we ought to have healthy suspicion about claims that suggest the East doesn't have a doctrine of the Fall (George Murphy surely overstates the difference in "Roads to Paradise and Perdition: Christ, Evolution, and Original Sin," *Perspectives on Science and Christian Faith* 58 [2006]: 109–18), or that the "Irenaean" account of origins somehow avoids any account of a "historical" Fall. For an incisive critique of the latter sort of claim, see Andrew M. McCoy, "Becoming Who We Are Supposed to Be: An Evaluation of Schneider's Use of Christian Theology in Conversation with Recent Genetic Science," *Calvin Theological Journal* 49 (2014): 63–84.

evolutionary accounts of human origins. This was already noted twenty-five years ago by John Polkinghorne.[3] More recently, scholars have concluded that, in fact, the traditional understanding of the Fall and original sin are simply incompatible. John Schneider, for example, has claimed that "matters of conflict between genomic evolutionary science and Christianity's standard western teaching on origins . . . cannot be resolved *hermeneutically*, but can only be resolved *theologically*, i.e., by revising what has become the quasi-orthodox Augustinian *theology* of origins as enshrined in Protestant confessions, as embedded in Protestant systematic theology, and as employed at crucial points in important Christian theodicy. At the core of this theology of origins is the doctrine of a historical Fall."[4]

In this chapter, I will argue that such claims are hasty at best, and more likely mistaken, resulting from a lack of theological imagination and a failure to appreciate just what is at stake in the traditional doctrine of the Fall. This requires exploration around two clusters of questions.

First, what's *at stake* in the doctrine of the Fall? Is the Fall just a myth to get at some insight about "the human condition," a received story to get at sinfulness—"original sin"—rather than an account of the origin *of* sin? The "mythical" reformulation of original sin has been a dominant position in current engagements between Christian theology and evolutionary accounts. But I suggest that this fails to recognize other theological commitments that are at stake in the doctrine of original sin and the origin of sin in a historical Fall. In particular, I will argue that what's at issue in the traditional doctrine of the Fall is not just an account of our "sinfulness" but also an account of the origin *of* sin, particularly when coupled with the historic doctrine of creation *ex nihilo*. The orthodox doctrine of the Fall, in other words, isn't simply a mythological account to explain what's wrong with humanity; it is an account of how that could be the case *if* creation *ex nihilo* is the case,

---

3. The Fall, says Polkinghorne, is "the major Christian doctrine that I find most difficult to reconcile with scientific thought." See *Reason and Reality: The Relationship between Science and Theology* (London: SPCK, 1991), 99.

4. John R. Schneider, "Recent Genetic Science and Christian Theology on Human Origins: An 'Aesthetic Supralapsarianism,'" *Perspectives on Science and Christian Faith* 62 (2010): 201, emphases original. As sources cited in fn. 2 above indicate, this theology of origins is not *only* "Protestant." Schneider appeals to an "Eastern" alternative in Irenaeus, but as others have demonstrated, Schneider's "Irenaeus" looks more like John Hick than the church father. See Andy M. McCoy, "Becoming Who We Are Supposed to Be," 63–84, and Mark S. M. Scott, "Suffering and Soul-Making: Rethinking John Hick's Theodicy," *Journal of Religion* 90 (2010): 313–34.

offering a theological account of human origins that doesn't jeopardize the goodness of God or human responsibility. I will suggest that there are also *eschatological* issues at stake here, insofar as any model that simply makes sinfulness synonymous with our "nature" will then have trouble accounting for what redemption and consummation look like in the eschaton (i.e., such models will be in danger of positioning the order of redemption *in opposition to* the order of creation).

Second, this project also requires some thinking about the nature of "history" in any notion of a "historical" Fall. On the traditional understanding of the Fall, something *happens*, and any such "event" would have to be something that happens in space and time. This issue of historicity—coupled with the fact that the traditional doctrine of the Fall does not rest (entirely) on Genesis 3—suggests that the conversation between evolution and Christian theology needs to grapple with ongoing discussions around the theological interpretation of scripture.[5]

I will first map and clarify the issues that are at stake in the traditional understanding of the Fall and then, as part of proposing an alternative model, I will take up questions related to the historical nature of the Fall and related questions of historicity.

What's at Stake? Original Sin and the Goodness of God

Narrative Context: The "Plot" of Creation, Sin, and Redemption

The doctrine of original sin is not a discrete proposition distilled from a single proof text, nor is the historical understanding of the Fall merely the byproduct of a particular ("literalist") reading of Genesis 3. Christian theology isn't like a Jenga game, an assemblage of propositional claims of which we try and see which can be removed without affecting the tower. Rather, Christian doctrine is more like the grammar of a story held together by the drama of a plot. In that sense, the doctrine of original sin and the historical understanding of the Fall is woven into the fabric of a story that is ultimately the drama of God's gracious interaction with humanity.

Following the lead of Alasdair MacIntyre and George Lindbeck, we might say that Christian theology is the distillation of the grammar of the Christian tradition—which is itself the outworking of a historical, covenant

---

5. See the chapters by Richard Middleton and Joel Green in this volume.

relationship between God and his people now entrusted to the church in the canon of scripture. The dramatic action of God in history is reflected in the narrative arc of scripture, which is in turn distilled in the catholic heritage of Christian theology (in a constellation of creeds, confessions, and guiding doctrinal authorities like Augustine, Aquinas, and Calvin, along with the liturgical heritage of the church's worship and spiritual disciplines[6]). The grammar of Christian theology encapsulates the biblical narrative in a *plot* that begins with the goodness of creation, a fall into sin, redemption of all things in Christ, and the eschatological consummation of all things.[7] The specifics or mechanics of this overall plot have been, and continue to be, understood differently across Christian traditions, but this narrative arc is, I take it, a "catholic" reading of scripture. The young earth creationist and the evolutionary creationist may understand the concrete specifics differently but still share a sense of this overall plot as both a reflection of the biblical witness and an encapsulation of God's relationship to his creation.

As we work through challenging questions at the intersection of human origins and evolution, the church is grappling with new questions about how coherently and responsibly to articulate this story. This has generated what Cardinal Newman has taught us to call "developments" in theology, new proposals *within* the Christian tradition. Utilizing MacIntyre's framework, I suggest the principal way a community discerns whether theological developments are "faithful extensions" of the tradition is by determining whether such developments are consistent with this core "plot." Furthermore, emplotment indicates that there is a kind of "narrative logic"—some moves "make sense" within a plot; other moves are puzzling and contradict the narrative thread of the story.[8] So if we are going to evaluate developments and proposals regarding human origins and original sin as to whether or not

---

6. I'm trying to describe this in a way that resonates with what Billy Abraham calls "canonical theism." See William J. Abraham, *Crossing the Threshold of Divine Revelation* (Grand Rapids: Eerdmans, 2006), 14–18, further developed in William Abraham, Jason Vickers, and Natalie B. Van Kirk, eds., *Canonical Theism: A Proposal for Theology and Church* (Grand Rapids: Eerdmans, 2008). For a discussion, see James K. A. Smith, "Epistemology for the Rest of Us: Hints of a Paradigm Shift in Abraham's *Crossing the Threshold*," *Philosophia Christi* 10 (2008): 353–61.

7. See, for example, J. Richard Middleton, *A New Heaven and a New Earth: Recovering Biblical Eschatology* (Grand Rapids: Baker Academic, 2014). I take it that this plot is "catholic" in the best sense of the term.

8. Consider Paul Ricoeur's discussion of "emplotment" in *Time and Narrative*, vol. 1, trans. Kathleen McLaughlin and David Pellauer (Chicago: University of Chicago Press, 1990), 41–54.

they are "faithful extensions" of the tradition, we will need to evaluate them against this fundamental plot.[9]

If we zoom in on this sweeping biblical narrative's account of sin and its origin, we see certain aspects of this plot shared across the historic Christian tradition—an Augustinian inheritance you can hear with different accents in the Catechism of the Catholic Church as well as the Heidelberg Catechism. Acknowledging my own particular location in the stream of catholic Christianity, let me use the sources of the Reformed confessions to distill three key, shared features (while also noting overlap with the Roman Catholic tradition):

The Goodness of Creation

The orthodox tradition emphasizes the original goodness of creation as a reflection of the goodness of the Creator.[10] More specifically, the confessions tend to focus on the original goodness of humanity. So when the Heidelberg Catechism poses questions about the origin of sin, it emphasizes the original goodness of humans created in the image of God: "God created [humanity] good and in his own image, that is, in true righteousness and holiness."[11] While we are concerned with the doctrine of original sin, in the confessions this is actually bound up with a picture of original righteousness: there is an origin *to* sin precisely because humanity was created in the image of God, "good, righteous, and holy."[12]

The confessions regard this as crucial for two reasons: First, it is essential to the biblical account that *goodness is prior to evil*, which informs the doc-

---

9. And the whole reason for articulating a core "plot" is precisely in order to have a criterion in place for determining what is essential and what is incidental. The idea is that if you discern the plot, and then change some part of *that*, then you're telling a different story. One can tell the story of David and Bathsheba, and the details of whether she was on a roof, in a garden, or on a balcony are not essential. But if one retells the story in such a way that there's no adultery, no Uriah, no Nathan declaring, "You are the man!"—well, then one is telling a different story. This isn't just a new rendition of the *same* story; it is a different tale.

10. Belgic Confession 13: "We believe that this good God, after he created all things, did not abandon them to chance or fortune but leads and governs them according to his holy will . . ." This is echoed in the *Catechism of the Catholic Church* in which the teaching on the Fall opens with the claim: "God is infinitely good and all his works are good" (§385).

11. Heidelberg Catechism Q&A 6; Belgic Confession 14.

12. Belgic Confession 14. Again, we find a similar concern in the *Catechism of the Catholic Church*: "original sin . . . is a deprivation of original holiness and justice" (§405).

trine of creation *ex nihilo*.[13] It is precisely this claim that distinguishes biblical, Christian faith from all sorts of pagan competitors. Second, the teaching of the goodness of creation is intended to preserve the goodness of God. Thus in Q&A 6 the Catechism explicitly rejects any notion that *God* created a corrupt, wicked, and perverse humanity. ("Q. Did God create people so wicked and perverse? A. No.") The Belgic Confession affirms the same: if we are going to affirm that "this good God . . . created all things," and yet also affirm that "God is not the author of, nor can be charged with, the sin that occurs," then it must be affirmed that creation is originally *good*.[14] Thus the confessions emphasize not just the goodness of the Creator (who precedes and is independent of the creation), but also the original goodness of the creation that comes from his hand, in particular the original goodness of humanity created in God's image.

The Irruption of Sin

Then whence sin? The Reformed confessions are unanimous in emphasizing that sin *befalls* a good creation—it is an irruption in the order of a good creation. Sin is not "natural" or some natural outgrowth of creation, nor was it intended for creation—though it was clearly a *possibility* that attended creation. But it is a fall *into* sin, a result of "the fall and disobedience of our first parents, Adam and Eve, in paradise" (HC Q&A 7). Man[15] "subjected himself willingly to sin and consequently to death and the curse" because he "transgressed the commandment of life, which he had received" (BC §14). This transgression and rebellion have consequences: separation from God (BC §14); the corruption of our nature such that humanity cannot *not* sin (BC §14–15); subjection to physical and spiritual death (BC §14); and the loss of "excellent gifts" including both free will and clear knowledge (BC §14).

Entailed by the coupling of these two claims—the original goodness of creation from the hand of a good God and the corruption of creation

13. Belgic Confession 12.

14. Belgic Confession 13. Below we will note that "goodness" should not be equated with "perfection." NB Heidelberg Catechism Q&A 115: ". . . until after this life we reach our goal: perfection."

15. Unfortunately, the confessions are not gender neutral, creating grammatical challenges for citation if we try to formulate as "humanity" rather than "man." We are simply conceding to the translation of the confessions in this formulation.

by human action—seems to be the "event character"[16] of the Fall. In other words, the confessions view the fall into sin and the irruption of fallenness as a historical, temporal reality.[17]

A Gracious Redemption and Eschatological Consummation

In some ways, the dynamics of creation and Fall are the stage for the good news of the gospel: the divine initiative of grace that rescues all of creation from its condition of brokenness and sin. "We believe that our good God, by his marvelous wisdom and goodness, seeing that man had plunged himself in this manner into both physical and spiritual death and made himself completely miserable, set out to find him, though man, trembling all over, was fleeing from him" (BC §17). This grace and mercy takes flesh in the Son who lives out what humanity was made for (BC §18–21; HC Q/A 12–19). Salvation is God's work, by God's gracious initiative, growing out of God's love, toward a restoration of all of creation that finds its fulfillment in an eschatological consummation of all things.[18]

This plot of creation, fall, redemption, and consummation constitutes the "core" plot of biblical faith. Any developments and proposals for "extending" the tradition will need to be accountable to this core in order to justify themselves as *faithful* extensions of the tradition. In that light, I will now consider some specific challenges and proposals related to the evolutionary

16. This is language used in synodical decisions of the Christian Reformed Church. Cp., for example, Report 28: "there are strict limitations on the extent to which the Genesis text can be reinterpreted within the Reformed tradition. However stylized, literary, or symbolic stories of Genesis may be, they are clearly meant to refer to real events. Especially in the case of God's acts of creation, Adam and Eve as first parents, the fall of humanity into sin, and the giving of the so-called 'mother promise' (Gen. 3:15), the reality of the events described is of foundational importance for the entire history of redemption" (403). This was reaffirmed by Synod 2010 (*Acts of Synod* 2010, xx–xxi).

17. As does the *Catechism of the Catholic Church*: "The account of the fall in Genesis 3 uses figurative language, but affirms a primeval event, a deed that took place" (§390). As I will argue in the next section, this need not entail a "literalist" reading of Genesis 1–3 as if the early chapters of Genesis offer some kind of *documentary chronology*. Nonetheless, it affirms that the *theological* teaching is about a *historical* reality.

18. Again, this resonates with the *Catechism of the Catholic Church*: "The doctrine of original sin is, so to speak, the 'reverse side' of the Good News that Jesus is the Savior of all men, that all need salvation and that salvation is offered to all through Christ. The Church, which has the mind of Christ, knows very well that we cannot tamper with the revelation of original sin without undermining the mystery of Christ" (§389).

understanding of human origins in order to consider how (or whether) they are "faithful" developments of this confessional tradition.

Evaluating Proposals and Developments: Faithful to the Plot?

Given the narrative parameters of this core plot, let's consider a few specific issues and proposals that have emerged in Christian theology's encounter with evolutionary accounts of human origins.

1. Does faithfulness to this plot require the affirmation of one historical couple as the origin of all human beings today?

Certainly historic Christian confessions refer to Adam and Eve and to their act of disobedience. And certainly the authors of the Reformed confessions straightforwardly believed in the existence of a historical couple. Continuing to affirm this is, in some sense, an understandable default position for Christian faith, and I am not advocating that we *a priori* rule out such a view in the name of academic respectability. However, the affirmation of a historical pair is significantly "cross-pressured"[19] in our current context. In light of accumulating archeological and genetic evidence, it is difficult today to *simply* affirm the existence of an original human couple, Adam and Eve. Indeed, such an affirmation entails a unique *theological* challenge: If all humans are descended from a single pair, why would the Creator of the universe seem to indicate in his creation (i.e., via general revelation) that humanity has a long, evolutionary origin and is descended from many more individuals? Any assertion of this received account of one historical couple will have to grapple not only with the scientific evidence to the contrary, but also with the theological problem that is generated when the "book of nature" seems to say something very different. There may indeed be theologically cogent ways to address this discrepancy, but it is important that we concede that the "traditional" picture of one historical couple, Adam and Eve, is not theologically unproblematic.

Given the theological challenges for what might seem the straightforward or "traditional" model, we can ask: Is the affirmation of a historical pair *required* for confessional faithfulness? Or could this be considered incidental

19. In *A Secular Age* (Cambridge, MA: Harvard University Press, 2007), Charles Taylor talks about the "cross-pressures" experienced by those who inhabit pluralistic contexts of competing plausibility structures (300–302).

to the central theological teachings about the creation of humanity in God's image? As Christian scholars are cross-pressured by evidence from the natural sciences about the history of the human species, could we imagine that the core plot aspect of the Fall could be preserved even in scenarios that entertain an evolutionary history of the human race?

Without prejudging the question of a historical couple, my project is more a thought project meant to stretch and prompt our theological imagination. I suggest there could be scenarios that both recognize a larger initial population *and* still affirm that God has created humanity in his own image, as recipients of revelation called into a relationship with God. These are difficult issues that will require a generation of interdisciplinary research and Spirit-led, scripturally faithful developments in our theological imagination—and it is important to create the *space* for the community to have *time* to conduct such research. The narrative constraint on such research is to account for scripture's *teaching* about the goodness of creation, the uniqueness of humanity, and humanity's original righteousness and rebellion into sinfulness. But it may be possible to account for those central affirmations without requiring one historical couple as the origin of all humans today.

2. Does faithfulness to this plot require that humanity was originally "perfect"?

Revisionists who have already concluded that an evolutionary account of human origins precludes a traditional understanding of the Fall point to the predatory ancestry of *Homo sapiens* as evidence that requires jettisoning any notion that humanity was "perfect" before the Fall.[20] This seems usually to assume a kind of genetic determinism: humanity has inherited DNA configurations from prehuman species that were violent and predatory, ergo humanity emerges inclined to the same. This presumes a category mistake, it seems to me, but I won't dwell on the point here.[21]

Instead, I would simply note that the confessions' affirmation of original

20. See, for example, Schneider, cited above. We should note that this particular question is distinct from a broader question of how one accounts for *death* (and hence) predation before the Fall, especially if—per the tradition—death (and hence the "evil" of predation) is taken to be a *consequence* of the Fall. This question lies outside the scope of this chapter. Suffice it to say that this might require redescribing pre-Fall death and reserving "violence" as a descriptor of human and post-human forms of predation. See Michael Murray, *Nature Red in Tooth and Claw: Theism and the Problem of Animal Suffering* (Oxford: Oxford University Press, 2011).

21. Cf. Sarah Coakley, "Evolution, Cooperation, and Divine Providence," in *Evolution,*

goodness and righteousness is not the equivalent of moral *perfection*. If pre-fall humanity was perfect, how could there have been a fall? In this respect, the Augustinian understanding of the Fall already recognizes the limits of original humanity's holiness. While humanity was created "good, just, and holy," as the Belgic Confession puts it, nonetheless humanity "did not understand it and did not recognize [its] excellence."[22] Perfection is only an eschatological reality in light of the grace of Christ: we are to pray "to be renewed more and more after God's image, until after this life we reach our goal: perfection."[23]

Thus the historical teaching regarding the Fall and original sin already recognizes certain limitations of even an originally "good" humanity. This leaves room for an appreciation of a kind of moral immaturity in original humanity—something that Augustine already recognized by noting the difference between the state of created humanity (*posse non peccare et mori*, able not to sin and die) versus redeemed humanity in the eschaton (*non posse peccare et mori*, not able to sin and die). So affirming the historical understanding of the original goodness of humanity is not equivalent to affirming an original *perfection* of humanity. Furthermore, an affirmation of the original goodness of our human forebears is not inconsistent with recognizing the need for moral growth and maturation.

3. Does faithfulness to this plot require that the Fall is a temporal, historical "event"?

Most traditional accounts of the Fall picture one "event"[24] in which Adam and Eve made *a* decision and fell into sin. Is that punctiliar[25] understanding of the Fall an essential *teaching* of the historical doctrine, or might it be regarded as a kind of incidental formulation? Indeed, must we subscribe to a temporal, historical notion of the Fall as an "event"[26]—as a change *in time*?

Games, and God: The Principle of Cooperation, ed. Martin A. Nowak and Sarah Coakley (Cambridge, MA: Harvard University Press, 2013), 375–86.

22. Belgic Confession 14.

23. Heidelberg Catechism Q&A 115.

24. Though one could point out that there are issues in the ballpark here even for the most "traditional" literalist view. Just *when* did the Fall "happen"? When *Eve* ate the fruit? When Adam did? When God issued punishment? How much "time" elapsed between them? This is not to encourage unfettered speculation; it is only to point out that even what might seem to be the most "conservative" account would need to grapple with issues of temporality.

25. That is, taking place at a *point* in time.

26. In my own ecclesiastical tradition, for example, Report 28 (1991) recognizes the unique genre of Genesis 1–11 as "highly stylized and compressed"; the account "does not

Here we hit upon one of the most difficult issues in current discussions. Certainly the historical, orthodox doctrine of the Fall as the *origin* of sin includes a picture of a temporal Fall as a discrete act—a fall *in time* from an original goodness to a sinful state, with repercussions for the entirety of creation, all as a result of the choices of an original pair, Adam and Eve (Rom. 8:20-21). However, this straightforward picture of a punctiliar event faces serious challenges from evolutionary scenarios. And any attempt to secure this traditional model will have to deal with the theological problem of apparent false history. In other words, while a certain burden of proof is borne by theological developments that depart from this traditional picture, the traditional picture should not get a "free pass," as it were: the assertion of one historical couple and a punctiliar Fall face theological challenges if we—for *theological* reasons—are going to take the science seriously.

In order to understand and evaluate these proposals, we need to carefully parse what's at stake in the doctrine of the Fall as received by the (Western) tradition. This requires recognizing the Augustinian heritage of such understandings of the Fall and original sin.[27] The Augustinian doctrine of the Fall and original sin was itself sort of "cross-pressured," formulated as a response to at least two heretical temptations: First and most importantly, Augustine was resisting Pelagian tendencies that would minimize the unmerited grace of God by suggesting any sort of "inherent" human ability with respect to salvation. For Augustine, the utter necessity of God's grace for salvation was paramount. Second, Augustine contested any teachings that would denigrate the goodness of creation (as the Manichees did) or attribute evil to a good God. At stake on

necessarily follow chronological order" (403, recommendation I). However, the Report then immediately notes: "there are strict limitations on the extent to which the Genesis text can be reinterpreted within the Reformed tradition. However stylized, literary, or symbolic these stories of Genesis may be, they are clearly meant to refer to real events. . . . Any interpretation which calls into question the event character of the story told in these first and fundamental chapters of the Bible must be firmly rejected, whatever difficulties this may cause with respect to scientific evidence" (403-4, recommendation J). It should be noted that Synod 2010 explicitly reaffirmed recommendation J (*Acts of Synod* 2010, 872-73).

27. While first formulated by St. Augustine, the doctrines of the Fall and original sin were not taken to be *invented* by Augustine, however. Rather, they are the systematic fruit of an engagement with the scriptures and seen as entailments from other core doctrinal commitments. Augustine, Luther, and Calvin all understood the "Augustinian" doctrine to be a Pauline doctrine and the teaching of the scriptures. I'm also not convinced that traditions that pride themselves on being non- or even anti-Augustinian (Orthodox, Wesleyans) thereby evade these same problems since most evasions end up "ontologizing" the Fall, making nature fallen *qua* nature, which then generates the theological problem of God's goodness.

this front was the very nature of God. As a result, the Augustinian doctrine of the Fall and original sin has at least these two aspects:

(a) An affirmation that goodness precedes evil—and more specifically, that humanity was originally righteous before rebelling and falling into sin. The priority of this "goodness" is not only logical and theological, but also chronological: humanity is created "good" and then temporally "falls."[28] Let's call this the "priority-of-the-good" thesis.

(b) A radical account of humanity as sinful, incapable of willing the good, and hence the necessity of an equally radical and unmerited grace as an action of divine initiative for redemption. Let's call this the "necessity-of-grace" thesis.

It is the priority-of-the-good thesis that seems to be challenged by evolutionary evidence regarding inherited dispositions. Thus some have tried to suggest that they are retaining an Augustinian or Reformed view simply by retaining the necessity-of-grace thesis.

But for Augustine, and the catholic Christian tradition, these were (and are) a "package deal." So the questions are: Can one take seriously the picture of human origins that emerges from evolutionary evidence and still affirm this Augustinian *package*? Or conversely: Can the priority-of-the-good thesis be entertained as nonessential? Or would tugging that thread unravel the entire fabric of the story?

My suggestion is twofold: that the Augustinian *package* is essential precisely because it is integral to the plot of the scriptures as disclosed in the grammar of Christian theology *and* that we could imagine affirming the evolutionary picture of a larger human population at the origin and still affirm both theses. This would depart from the *punctiliar* aspect of the traditional model but would still retain a *temporal, historical* understanding of the Fall. There could be different ways of imagining this. For example, some proposals imagine a group of individuals, selected by God to represent the rest of humanity, receiving a special revelation and commissioning, "good" (though not "perfect") and able to obey God's just requirements, who through acts of disobedience over a discrete period of time (with a clear "before" and "after") fell into a state of sinful rebellion from God.

---

28. This is not just an ontological affirmation that good precedes evil in the sense that the goodness *of God* is prior to and more fundamental than evil; it is the stronger claim that humanity was originally good.

Furthermore, there is a theological (and perhaps just plain logical) incoherence to any proposal that wants to retain the necessity-of-grace thesis without the priority-of-the-good thesis. For example, if someone rejects the (chronological) priority-of-the-good, then it would appear that God is to blame for human sinfulness since humanity would have emerged *as* sinful. In that case, it's not clear that God's redemptive activity is "unmerited." Indeed, on such models it sometimes sounds as if God is *obligated* to save—which undercuts the gratuity of God's grace. Furthermore, such models seem to run afoul of making the order of redemption *contra* the order of creation; that is, redemption (and eschatological consummation) would seem to require *undoing* rather than restoring creation. Grace ends up *contra* nature.

This is why, I believe, the inherited, traditional understanding of the Fall and original sin seems to require the *temporality* of the Fall—that sin is the outcome of human rebellion, issuing in a fall from a good creation to a sinful state. However, it does not appear that a punctiliar model with cosmic effects is required to retain the essential doctrine that the Fall is temporal, historical, and the result of human rebellion. The teaching that the Fall is historical and temporal (has an "event-character," as my denomination's synodical statements have put it) does not seem to require that it was necessarily punctiliar. Thus any models or scenarios that honor this temporal and historical nature of the Fall, even while demurring on the picture of one specific event, could be entertained as consistent with the core teaching of the confessions.

It should be noted that this does *not* require reading Genesis 1–3 "literally"—if, by literally, we mean taking "day," for instance, to refer to a twenty-four-hour period.[29] Our options are not *either* ahistorical "theological" claims *or* literalist "historical" claims. We shouldn't confuse or reduce "historical" to mean something like a blow-by-blow chronology recorded by CNN. We need to develop more nuanced accounts of history in order to do justice to the theological, which is precisely why discussions at the intersection of theology and evolutionary sciences need to take up hermeneutical questions at play in contemporary debates about the theological interpretation of scripture.[30]

---

29. This, of course, is not the historic understanding of "literal." For helpful discussion, see Henri de Lubac, *Medieval Exegesis*, vol. 2: *The Four Senses of Scripture*, trans. E. M. Macierowski (Grand Rapids: Eerdmans, 2000).

30. This lies beyond the scope of the present chapter. For a seminal discussion, see Joel B. Green, "Rethinking 'History' for Theological Interpretation," *Journal of Theological Interpretation* 5 (2011): 159–74. See also Richard Hays's incisive points in "Knowing Jesus: Story, History, and the Question of Truth," in *Jesus, Paul, and the People of God: A Theological Dialogue with N. T. Wright* (Downers Grove, IL: InterVarsity, 2011), 41–61.

Conclusion: A Modest Proposal

In conclusion, let me try a little thought project. The question I have tried to pose is this: Does taking evolutionary and genetic evidence seriously *require* that we revise and reformulate the doctrines of original sin and the Fall to the point of eliminating any historical, episodic aspect to the Fall? Or could we imagine models that both honor accumulating evidence of ancestral evolution *and* take seriously the tradition's understanding of the Fall and original sin? I think we can. So, for the sake of argument, let me float a provisional model as a kind of imaginative experiment. Allow me to paint a possible model to show how affirming genetic and evolutionary evidence of common ancestry does not necessarily entail abandoning a historical understanding of the Fall. I think such a model navigates through a number of the challenges I've noted above. However, I don't want to defend this sketch as the last word. Rather, I float the scenario as one open to correcting and refinement.

What if we imagined a scenario something like this:

> *In the beginning, God created the heavens and earth. From what he seems to tell us via the book of nature, the mechanics of creational unfolding was an evolutionary process: the emergence of new life was governed by the survival of the fittest, such that biological death and animal predation are part of this process, even part of what can be acclaimed as a "good" creation. So some of the phenomena we might have traditionally described as "outcomes" of the Fall seem to be part of the fabric of a good, emerging creation.*
>
> *From out of this process there emerges a population of hominids who have evolved as cultural animals with emerging social systems, and it is this early population (of, say, 10,000) that constitutes our early ancestors. When such a population has evolved to the point of exhibiting features of emergent consciousness, relational aptitude, and mechanisms of will—in short, when these hominids have evolved to the point of exhibiting moral capabilities—our creating God "elects" this population as his covenant people. The "creation" of humanity, on this picture, is the first election—the first of many (Noah, Abraham, Jacob, et al.). And in that covenantal election of a population, Yahweh established a relationship with humanity that involved his self-revelation to them and established moral parameters for them and for their flourishing. In being so elected, these pinnacle creatures are also deputized and commissioned as God's "image bearers"—the creator's representatives to and for creation's care and flourishing. They are charged with unfurling the latent potential enfolded into creation. And to some extent, creation now depends*

on their care and cultivation such that, should this emergent humanity fail to carry out its mission and obligations as articulated by God's "law," there will be "cosmic" consequences.

This original humanity is not perfect (the catholic theological tradition has never claimed that). They are able to carry out this mission—God's law would not be established where obedience is not possible—but they are also characterized by moral immaturity, since moral virtue requires habituation and formation, requires time. So while they are able to carry out this mission, there are no guarantees, and also no surprises when they fail. Since we're dealing with a larger population in this "garden," so to speak, there is not one discrete event at time T1 where "the transgression" occurs. However, there is still a temporal, episodic nature of a Fall. We might imagine a Fall-in-process, a sort of probationary period in which God is watching (not unlike the dynamics of the flood narrative in Genesis 6, a kind of second Fall narrative in the Torah). So the Fall might take place over time T1–T3. But there is some significant sense of before and after in this scenario.

And things change in the "after": there are cosmic effects of some discernible nature (cp. Colossians 1–2); there is also the cosmic fallout of humanity's failure to cultivate and care for creation; and there is also some kind of (almost?) ontological shift in human nature, or at least a certain solidification of human character in a certain direction and tendency that will require the regenerating initiative of God to make rightly ordered virtue a possibility. But this regeneration and sanctification will not constitute an undoing of their created tendencies and capacities, but rather a restoration[31] of creational possibilities and empowerment/formation to be able to realize that calling. Redemption will also require a grace that is cosmic in scope, a grace that is the outcome of the cross (Col. 1:20).

Now, for some this might already look like a reformulation that has abandoned the tradition; some might judge this not as a faithful extension of the

---

31. Actually, redemption will ultimately have to be *more* than mere restoration, otherwise the Fall would still be a possibility. This is exactly why Augustine emphasized that original humanity "in the garden" was *posse non peccare* (able not to sin) whereas eschatological humanity "in the kingdom" will be *non posse peccare* (not able to sin). So redemption is restoration but also *more* than restoration. Jon Stanley has carefully pointed out how Herman Bavinck honors this "something more" dynamic of redemption in ways that Kuyper does not; so Stanley describes Bavinck's account of redemption as "restoration *plus.*" See Stanley, "Grace Restores and Renews Nature," *Kuyper Center Review,* vol. 2: *Revelation and Common Grace,* ed. John Bowlin (Grand Rapids: Eerdmans, 2011).

tradition but rather a compromising concession to "science"—one that has given up on a historical couple and a split-second Fall as the result of a single decision, etc. But I can also imagine how one could consider those particular features of the received doctrine as nonessential.

However, this scenario does remain committed to other features of the traditional doctrine. Let me highlight just two:

First, on this scenario, the Fall is still historical, temporal, and even "eventual," though it is something like an episode-in-process.[32] This also retains a sense of cosmic effects as part of an "after." Such a historical picture seems to be required in order to retain a sense of sin as "not the way it's supposed to be." And *that* seems essential to the tradition to me, such that any faithful extension of the tradition will need to articulate continuity with that sensibility.

Second, this model resists "ontologizing" the Fall. I think an important point of concern with some recent proposals for reformulation is the way that they would end up "naturalizing" sin—inscribing brokenness into the fabric of creation (such that describing this as a "symbolic fallenness" is really a bit of a ruse since there's really no Fall *from* in this picture). There are important theological concerns at stake here, concerns that Christians have often discussed in terms of the relationship between nature and grace. While Protestants and Catholics have sometimes disagreed on their relation (grace "restores" nature *vs.* grace "perfects" nature), they were all agreed in their opposition to any notion that grace *opposes* nature or "nullifies" nature or "undoes" nature. In other words, the Christian tradition has consistently judged that any construal of grace/redemption that sees redemption as somehow anti-nature would be *ipso facto* anti-creation, and thus would posit a fundamental inconsistency within the biblical narrative. Making sin original is *not* the doctrine of original sin; it is a version of Gnosticism.[33]

My goal here has been to map the terrain of questions and challenges,

32. I think we make room for something like this in other contexts. For instance, when did I "win" the Daytona 500? Only at the checkered flag? What if I was leading for the final twelve laps? Or when did I earn a gold medal for the marathon? Only when I crossed the finish line? The "event" of my "win" does not seem to be simply punctiliar. Every coach knows this when he points out that, while the other team beat us with a score as the clock ran out, we "lost" the game earlier by missing scoring chances, etc. The point is that our folk notion of an "episode" is quite elastic.

33. I have criticized Stephen Mulhall's *Philosophical Myths of the Fall* for just this point: ontologizing or naturalizing the Fall such that creation is always already fallen. See Smith, *The Devil Reads Derrida—and Other Essays on the University, the Church, Politics, and the Arts* (Grand Rapids: Eerdmans, 2009), chap. 12.

charting where and why I think there are boundaries for this conversation—
and what's at stake in crossing them. But I hope I've also indicated the room
that remains in those capacious parameters for creative, constructive, and
faithful responses to contemporary challenges. The chapters that follow will
extend and deepen these possibilities.

PART II

Biblical Studies and Theological Implications

4 Reading Genesis 3 Attentive to Human Evolution

Beyond Concordism and Non-Overlapping Magisteria

J. RICHARD MIDDLETON

Let us begin by reimagining the scripture/science conversation. Although there are divergences of opinion on details (since the science is always being refined), most paleo-anthropologists date the first hominin remains to some six or seven million years ago, with the Australopithecines appearing about four million years ago and the genus *Homo* about two million years ago (*Homo habilis*).[1] The most likely current hypothesis for the evolution of anatomically modern *Homo sapiens* places their origin some 200,000 years ago, with a minimum population of anywhere from 2,000 to 10,000.[2]

Many skeptics and committed Christians alike have judged this scientific account incompatible with the biblical version of the origin of the humanity recounted in the early chapters of Genesis. From the skeptical side, the Bible has often been dismissed because its mythical or prescientific account of origins (both cosmic and human) is thought to contradict what we know from modern science. This skeptical approach is most evident in the "warfare" model of science and religion made famous by John W. Draper and Andrew Dickson White in the nineteenth century, and perpetuated by the

1. Contemporary anthropologists have come to use the term *hominin* (rather than *hominid*) to refer to the grouping of humans with their prehuman relatives (this includes the genus *Homo*, as well as distant relatives, such as the *Australopithecines*). The term *hominid* now refers to the larger grouping, including all monkeys and apes.

2. Nothing in what follows is determined by these particular estimates; science is a fallible, ever-changing project, and it is to be expected that details of these estimates will be disputed, and indeed will change over time.

new atheists like Christopher Hitchens and Richard Dawkins in the twentieth and twenty-first centuries.[3]

Many Christians (especially evangelicals and fundamentalists in North America) have bought into the warfare model, with the difference that they assume the "literal" truth of the biblical account—taking "literal" in the sense of necessitating a one-to-one correspondence between details of this account and events and actualities in the empirical world.[4] This approach, which often goes by the name "scientific creationism" or "creation science" (or, more recently, "origin science") assumes that the Bible intends to teach a true scientific account of cosmic origins—including a young earth and the discontinuity of species (particularly the discontinuity of humans from other primates).[5]

Since this way of reading biblical creation accounts clearly contradicts the understanding of origins provided by modern science (both in cosmology and in evolutionary biology), proponents of "creation science" typically dismiss the putative claims of modern science (at least in the case of cosmic and biological origins) as ideologically tainted. The result is a concordist attempt to force science to fit what the Bible is thought to say about these topics.[6]

3. John W. Draper, *History of the Conflict between Religion and Science* (1874); Andrew Dickson White, *A History of the Warfare of Science with Theology in Christendom* (1896); this latter is an expanded version of an earlier and shorter work by White, titled *The Warfare of Science* (1876). Christopher Hitchens, *God Is Not Great: How Religion Poisons Everything* (New York: Twelve Books, 2007); Richard Dawkins, *The God Delusion* (Boston: Houghton Mifflin, 2006).

4. There is another sense of "literal," from the Latin *ad literatum*, equivalent to reading according to the intended genre of the work.

5. Both the skeptical and Christian assumption of a "warfare" model can be understood as versions of the "conflict" model of relating religion and science propounded by Ian Barbour in his famous fourfold typology of their possible relationships (Barbour, *Issues in Science and Religion*, first published in 1966).

6. A more recent concordist approach works in the opposite direction, attempting to harmonize the Bible with the conclusions of modern science. This approach, spearheaded by Hugh Ross and the organization called "Reasons to Believe," attempts to make the Bible agree with modern cosmology (the Reasons to Believe website is: http://www.reasons.org/). In this approach, the Bible's statements about the nature and origin of creation are not understood in their ancient conceptual context, but interpreted so as to make them harmonize (anachronistically) with modern scientific claims (including a universe of galaxies billions of years old). Yet at one point this concordist project agrees with that of "creation science"—biological evolution (especially human evolution) is beyond the pale. See, for example, Fazale Rana and Hugh Ross, *Who Was Adam? A Creation Model Approach to the*

One of the most problematic dimensions of affirming both biblical origins and biological evolution is the doctrine of the "Fall," since the Bible seems to teach (in Genesis 3) a punctiliar event in which an original couple transgressed God's commandment after an initial paradisiacal period. Whether the classical doctrine of "original sin" is required (in all its specificity) for creedal orthodoxy is an open question. Nevertheless, the Bible itself certainly seems, at first blush, to tie the origin of evil to an understanding of human beginnings that is quite different from what we find in evolutionary biology.

Given the putative contradiction between biblical-theological claims and evolutionary science, what's an honest Christian to do? Suppose someone wants to do justice both to biological evolution and to the historic Christian faith ("that was once for all entrusted to the saints"; Jude 3), how might one go about affirming both with integrity?[7]

The most common approach has been to utilize some version of Stephen Jay Gould's proposal of Non-Overlapping Magisteria (NOMA), which would separate biblical and theological truth from scientific truth as belonging to distinct conceptual domains, which therefore guarantees no contradiction between them.[8]

Variants of NOMA can be found, with or without the explicit terminol-

---

Origin of Man (Colorado Springs: NavPress, 2005). Among Ross's many books is his early *The Fingerprint of God* (Orange, CA: Promise Publishing, 1989; 3rd ed. 2005). For Ross's more recent attempt to harmonize science and the Bible, see *Hidden Treasures in the Book of Job: How the Oldest Book of the Bible Answers Today's Scientific Questions* (Grand Rapids: Baker, 2011). The advertising for this book states that "Job is filled with rich insight into both ancient and modern questions about the formation of the world, the difference between animals and humans, cosmology, dinosaurs and the fossil record, how to care for creation, and more."

7. From here on all biblical quotations will be from the New Revised Standard Version unless otherwise noted.

8. Stephen Jay Gould, "Non-Overlapping Magisteria," *Natural History* 106 (March 1997): 16–22. Gould proposed this way of conceiving the relationship of theology and science in the aftermath of Pope John Paul II's address on evolution and faith to the Pontifical Academy of Sciences in 1996, although he traces his reflections on the issue back to a 1984 trip to the Vatican City (sponsored by the same Academy) during which he discussed evolution and Christian theology with a group of Jesuit scientists. Gould makes the point that he is not inventing the approach of NOMA, only the terminology. Gould explains that this has been the *de facto* approach of the Catholic Church since at least Pope Pius XII's 1950 encyclical, *Humani generis*, and that John Paul's 1996 address was a self-conscious attempt to address a certain reticence on the part of Pius regarding the *factuality* of evolution (even though he had proposed that there was, *in principle*, no contradiction between evolution and faith).

ogy, among many writers on the subject of Christianity and evolution, since it provides a helpful methodological alternative to the warfare model.[9] In contrast to the assumption of many evangelical or fundamentalist Christians that an evolutionary account of human origins is incompatible with the biblical account of "Adam," increasing numbers of scientists and theologians today are attempting positively to affirm an orthodox Christian faith along with scientific findings about biological evolution. Whether described as "theistic evolution" (the older term) or "evolutionary creation" (the more recent term, used, for example, by BioLogos), this attempt to honor both the non-negotiable authority of scripture and the cumulative research of more than a century of paleontology, along with the recent contribution of genetics, is commendable.

As an alternative to a naïvely concordist attempt at reconciling scripture with science, the embrace of NOMA by contemporary Christians is fully understandable. It allows evolutionary scientists to get on with their work, without having to compromise their findings with the putative truths of theology. And theologians can likewise reflect on God's role in the biological processes of life's development, without being proscribed by science.

But is that all there is to be said? As a biblical scholar, am I to simply bracket the scientific account of human origins (and ignore what I know of hominin evolution) when I interpret Genesis 3? Certainly, the assumptions and presuppositions of the interpreter must affect—in some way—what he sees (and doesn't see) in scripture. And does the Bible not have any relevance for thinking about evolution? In what follows I intend to think evolution *together* with the biblical account of the origin of evil in Genesis 3.

Here I am emboldened by the work of Old Testament scholar William Brown, especially in his attempt to move beyond both concordism and NOMA to an exploration of possible "resonances" that might arise from a "cross-disciplinary conversation" between the Bible and science.[10]

In his brilliant and inspiring work *The Seven Pillars of Creation*, Brown explores the major creation texts in the Old Testament (including Genesis 2–3) in connection with contemporary science, utilizing a three-step method. Beginning with an *elucidation* of each text, Brown then *associates* the theological themes of the text in question with what he discerns might

9. NOMA seems to correspond to Ian Barbour's "independence" model of the relationship of religion and science. That is, there isn't any conflict between them, but the nature of the relationship is not clearly specified.

10. William P. Brown, *The Seven Pillars of Creation: The Bible, Science, and the Ecology of Wonder* (Oxford: Oxford University Press, 2010), 8.

be relevant aspects of the world we know from science, which he then explores. Finally, he then returns to the biblical text with the insights gained from science in order to *appropriate* the text for its wisdom and relevance for life today. Brown conceives this process as "a hermeneutical feedback loop"[11] between the biblical text and contemporary science whereby a variety of "consonances," "correlations," "connections," "points of contact," or "parallels" between the text and our scientific knowledge may be explored.[12]

What prevents this from simply being a new attempt at concordism or harmonization?[13] First, Brown is clear that these connections are "*virtual* parallels," "*analogous* points of contact or *imaginative* associations"— in other words, there is an ineluctable element of interpretive subjectivity here.[14] Second, Brown treats scripture as an ancient text, with no knowledge of contemporary science, and acknowledges that we therefore need to be aware of "claims made by the biblical text about the world that *conflict* with the findings of science"; he thus suggests that we attend to the "disjunctions" and "collisions" as much as to the resonances.[15] That this is different also from NOMA is clear, since on that model neither discourse, the biblical-theological nor the scientific, is allowed to inform the other. Thus Brown suggests (tongue in cheek) that we might think of his approach as "TOMA or 'tangentially overlapping magisteria.'"[16]

Is Brown then suggesting that contemporary science should shape our theology or our interpretation of scripture? Not quite. His suggestion is that while science should not dictate the direction of biblical interpretation, it may "nudge the work of biblical theology in directions it has not yet ventured and, in so doing, may add another layer to Scripture's interpretive 'thickness' . . . or wondrous depth."[17]

My approach to the relationship of scripture and science in this chapter is similar to that of Brown, with three caveats or differences. First, whereas Brown focuses on the relationship between creation texts and contemporary

11. Brown, *Seven Pillars of Creation*, 16.

12. Brown, *Seven Pillars of Creation*, 9–10.

13. In conversation, Brown has quipped that *completely* overlapping magisteria would result in COMA!

14. Brown, *Seven Pillars of Creation*, 10 (my emphases).

15. Brown, *Seven Pillars of Creation*, 10 (my emphasis).

16. Brown, *Seven Pillars of Creation*, 17. We should note that Gould himself admitted that the two domains of science and theology often rub up against each other in interesting ways, which require negotiation.

17. Brown, *Seven Pillars of Creation*, 16.

science, I will attempt to read the narrative of the "Fall" in Genesis 2–3 in relation to what we know of the evolutionary history of *Homo sapiens*. Second, whereas there are many dimensions of the scientific understanding of the world that Brown is able to draw upon in his interpretation of biblical creation accounts, there is very little that scientists understand about the origin of religion, morality, and ethics among *Homo sapiens*. Finally, whereas Brown is able to move in the scope of his lengthy book from the biblical text to contemporary science, and then back to the biblical text, the space limitations of this chapter preclude any such lengthy three-part exposition.

My approach will be to range over a number of prominent themes or motifs in the garden story of Genesis 2–3, exploring the significance of these themes for human evolution and, alternatively, how an understanding of evolution might help us interpret the themes or motifs in the texts (although sometimes I may simply raise questions to which I don't have clear answers at the moment). I thus conceive of this chapter as an experimental probe in two directions—to see if the biblical text might help us think about the origin of moral consciousness among *Homo sapiens* and whether our current knowledge of the evolution of *Homo sapiens* might illuminate aspects of the text that readers have previously missed. Along the way, my reading of the biblical text and the evolution of *Homo sapiens* will draw upon a virtue-ethics approach to the development of moral consciousness. My hunch is that a close reading of Genesis 2–3 in connection with human evolution might shed light on conceptualizing the origin of moral evil, including the notion of a "historical" or "eventful" Fall.[18]

The *'ādām–'ădāmâ* Connection

Although my focus in what follows will be on Genesis 3, this chapter is part of a larger, coherent literary unit that begins with Genesis 2:4b. It is, therefore, not inappropriate to begin with the origin of humanity as portrayed in Genesis 2.[19]

18. Thus I am working with the thesis articulated by James K. A. Smith (in his chapter in this volume) that some notion of a historical origin of human evil is consistent with (perhaps demanded by) orthodox Christian faith.

19. Much more could be said about the relationship between the Bible's depiction of human origins and what we know from the state of current evolutionary science. Recent works (with slightly different perspectives) include Peter Enns, *The Evolution of Adam: What the Bible Does and Doesn't Say about Human Origins* (Grand Rapids: Brazos, 2012);

Let us start with the name *Adam*. Is it significant that this name (like many of the names in the early chapters of Genesis) is clearly symbolic? Adam (*'ādām*) means "human." Indeed, Adam becomes a proper name only in Genesis 4 and 5; prior to that he is *hā'ādām* (the human).[20] So we seem to be justified in viewing him both as the first human and archetypally as everyman or everyone.

We should also note that the word for the first human (*'ādām*) functions as part of a Hebrew pun or wordplay throughout Genesis 2 and 3, where it sounds like (or resonates aurally with) the word for soil or ground (*'ădāmâ*). Biblical scholars have suggested various equivalent English puns, such as the *earth creature* from the *earth*, the *groundling* from the *ground*, the *human* from the *humus*.[21] The point is that the aural resonance of *'ādām* and *'ădāmâ* suggests a primal ontological resonance between the human and his earthly context. Not only is the human taken from the ground (a matter of derivation or origin), the human purpose is to work the ground (a matter of calling or vocation). Due to human sin, the ground is cursed, in the sense that the human's relationship with the ground becomes difficult (work becomes toil); primal resonance becomes dissonance. And death is described as returning to the ground from which the human was taken.[22]

Throughout this entire storyline, the aural resonance of human and ground (*'ādām* and *'ădāmâ*), along with the narrated contours of their interdependence, suggests that humans are fundamentally earth creatures or

---

and John H. Walton, *The Lost World of Adam and Eve: Genesis 2–3 and the Human Origins Debate* (Downers Grove, IL: IVP Academic, 2015).

20. There are four places in the narrative of Gen. 2–3 where *'ādām* appears without the definite article, but none of these is a proper name. According to 2:5, "there was no-one [lit. no *'ādām*] to till the ground." In Gen. 2:20, 3:37, and 3:21 we have *lĕ'ādām* (to/for the human); here the preposition *lĕ* (*to* or *for*) is appended to *'ādām* without the vowel change that usually indicates a definite article (*lā'ādām*). However, in the first case (2:20), the same verse also uses *hā'ādām* (the human); and it should be remembered that there would have been no distinction in the Hebrew consonantal text (so the Masoretic Text vowel pointing may be idiosyncratic). Gen. 4:25 is the first clear use of *'ādām* without the definite article ("Adam knew his wife again"). Yet Gen. 4:2, which first mentions the man knowing his wife, has *hā'ādām*. In Gen. 5:1, which begins a genealogy, we finally have the proper name *Adam* clearly intended.

21. For example, Phyllis Trible suggests the translation "earth creature" taken from the "earth," in *God and the Rhetoric of Sexuality*, Overtures to Biblical Theology (Philadelphia: Fortress, 1978), 76–78; Brown, *Seven Pillars of Creation*, 81–88.

22. For a fuller exploration of the centrality of the *'ādām–'ădāmâ* connection in the Primeval History, see Patrick D. Miller Jr., *Genesis 1–11: Studies in Structure and Theme* (Sheffield: JSOT, 1978), chap. 3: "The *'ădāmāh* Motif."

groundlings. This—together with the fact that the animals are also taken from the ground (Gen. 2:19)—may be helpful in thinking about how the picture of humanity in Genesis 2 might relate to what we know of human and animal origins from evolutionary history. Might this continuity of human and ground even help us in thinking about the similarity of many traits and abilities found in various animals that were once thought to be unique to humanity?

The Garden of Eden and the Breath of Life

In Genesis 2, the locale for primordial humanity is a garden. This garden, with its trees, rivers, and mention of precious and semiprecious stones, is reminiscent of a royal garden or sacred grove in the ancient Near East, a locale fraught with divine presence.[23] Whereas Genesis 1 draws on the conceptuality of heaven and earth as a cosmic temple, with humanity as God's "image" or cult statue in the temple, meant to mediate divine presence and rule from heaven to earth (heaven functioning as the cosmic Holy of Holies), the garden in Genesis 2 is the locus of divine presence on earth, where God "walks" in proximity to humanity.[24]

It is also significant that a sacred grove beside a primeval river is the typical setting for the *mīs pî* (mouth washing) or *pīt pî* (mouth opening) ritual, known from Mesopotamian texts. This was the ritual process through which a humanly constructed cult image was vivified and transformed ("transub-

23. See Gordon J. Wenham, "Sanctuary Symbolism in the Garden of Eden Story," *Proceedings of the World Congress of Jewish Studies* 9 (1986): 19–25. Reprinted in *I Studied Inscriptions from Before the Flood: Ancient Near Eastern, Literary, and Linguistic Approaches to Genesis 1–11*, ed. Richard S. Hess and David Toshio Tsumura, Sources for Biblical and Theological Study 4 (Winona Lake, IN: Eisenbrauns, 1994), 399–404.

24. Besides Wenham's pioneering work, see Gregory K. Beale, *The Temple and the Church's Mission: A Biblical Theology of the Dwelling Place of God*, New Studies in Biblical Theology 17 (Downers Grove, IL: IVP Academic, 2004); T. Desmond Alexander, *From Eden to the New Jerusalem: An Introduction to Biblical Theology* (Grand Rapids: Kregel, 2009); and the essays in *Heaven on Earth: The Temple in Biblical Theology*, ed. T. Desmond Alexander and Simon J. Gathercole (Carlisle, UK: Paternoster, 2004). I have addressed the motif of the cosmos as temple in *The Liberating Image: The Imago Dei in Genesis 1* (Grand Rapids: Brazos, 2005), chap. 2; "The Role of Human Beings in the Cosmic Temple: The Intersection of Worldviews in Psalms 8 and 104," *Canadian Theological Review* 2, no. 1 (2013): 44–58; and *A New Heaven and a New Earth: Reclaiming Biblical Eschatology* (Grand Rapids: Baker Academic, 2014), chaps. 2 and 8.

we know about the evolution of biological organisms, since mortality seems to be intrinsic to biological life. Even Genesis contradicts this interpretation when it portrays God forming the human from the dust of the ground, which is a metaphor for mortality—"you are dust, and to dust you shall return" (Gen. 3:17). Even Paul calls Adam a "man of dust," referring to his having been created mortal (1 Cor. 15:42–49).[38]

It is possible, however, that *death* could be taken as a *reversion* to mortality, assuming that the tree of life symbolizes immortality and that humans had eaten of its fruit prior to disobedience. However, the tree of life is more properly a symbol for earthly flourishing, in line with the wisdom literature of the Old Testament, which describes wisdom as a "tree of life" for those who find her (Prov. 3:18). This connection of wisdom with life is not only a pervasive theme in Proverbs (living according to wisdom leads to flourishing), but it might make sense of the garden story with its two trees, one of knowledge/wisdom and one of life.[39] This suggests a third meaning of *death*, namely as the antithesis of flourishing. So when the wisdom literature contrasts the two paths of Life and Death, this is not reducible to the contrast between mere existence and the extinction of existence; nor does it refer to immortality versus mortality. Rather, the focus is on the difference between a life that conforms to wisdom, rooted in reverence for God, which results in blessing and shalom, and a life of folly, characterized by rejecting God's ways, which is thereby deformed and plagued by corruption and calamity.[40]

It is this sense of *death* that allows the writer of Psalm 88 to claim that he is already in the grave (88:3–6). Death has begun to encroach on life; corruption has compromised normative flourishing. In a similar vein, when Jacob thought Joseph was dead, "he refused to be comforted, and said, 'No, I shall go down to Sheol to my son, mourning'" (Gen. 37:35). Jacob was not

38. Psalm 103:14 describes human mortality using the very words "formed" and "dust" from Genesis 2:7.

39. Since wisdom and life are associated elsewhere in the Bible, the question arises as to why they are separated into two trees in Genesis 2–3. This separation seems to serve the point of distinguishing (1) the initial, childlike wisdom that is equivalent to simply trusting God from (2) the mature wisdom that involves discerning between good and evil. The first sort of wisdom (appropriate to the initial stage of moral development) leads to life (it is compatible with eating from the tree of life); but the way in which one discerns good and evil may lead to life or death (thus exile from the garden). I will return to this distinction.

40. On the two paths, and the relation of wisdom to life, see Middleton, *A New Heaven and a New Earth*, chap. 5: "Earthly Flourishing in Law, Wisdom, and Prophecy."

planning suicide. Rather, the quality of his life had been compromised; life had become as death to him. This understanding ultimately leads to Paul regarding Sin and Death as powers (which stand in antithesis of life) that are overcome in the cross and resurrection of Christ.[41]

If we take the warning about death in Genesis 2 in this light, it not only coheres with the worldview of the rest of scripture, but it allows us to see mortality as an ordinary and even intrinsic component of the world God made. That organisms die, which is essential to evolutionary history, would not be in any sort of tension with the biblical accounts of creation.[42]

This does not mean that we should exclude immortality as the ultimate result of eating from the tree of life. After all, the reason that sinful humans are later exiled from the garden is because they might in their sinful state eat from the tree of life and "live forever" (Gen. 3:22). This allows us to see a canonical trajectory from the tree of life in Genesis 2 to its culmination in the New Jerusalem (Rev. 22:2, 14). In other words, it seems that God would have, at some point, after humans were *confirmed* in their obedience, made their flourishing (and the flourishing of the world) permanent. This interpretation draws on Paul's notion of the resurrection body as immortal or incorruptible (1 Cor. 15:50–54). As it turned out, however, the permanent flourishing of the world was disrupted by the intervention of sin, which would require a restorative act (redemption) to bring the world to its intended telos.

The Tree of the Knowledge of Good and Evil

There have been a number of divergent opinions in the history of interpretation about the significance of "the tree of the knowledge of good and evil," from which humans are commanded not to eat (Gen. 2:9, 17). Some interpreters appeal to the later narrative context, where the man "knew" his wife (Gen. 4:1; cf. 4:17, 25) and she conceived and bore a child, to suggest that it was sexual "knowledge" that was prohibited.[43] This may be taken in

---

41. See Beverly R. Gaventa, "The Cosmic Power of Sin in Paul's Letter to the Romans: Toward a Widescreen Edition," *Interpretation* 58, no. 3 (2004): 229–40.

42. Paul writes that the sting of death is sin (1 Cor. 15:56), which suggests that without sin death might not be regarded as an evil.

43. Thus we speak of "carnal knowledge" today. This reading sometimes appeals to Deut. 1:39, which speaks of children who don't yet know good and evil; but this may refer not to sexuality but to moral discernment (as is more likely, given other uses of this phrase in the Old Testament).

an Augustinian tone, which denigrates sex because of the lust involved, or in a more modern sense that eating of the tree was a fall "upwards" or "forwards" into maturity (which includes, but is not limited to, sex); the modern approach typically includes a tragic element in this fall upwards. One recent version of the fall into maturity suggests that the "knowledge of good and evil" refers to humanity coming to know from experience the struggle for existence, which includes suffering as part of the growth process (the Hebrew word *ra'* is not limited to moral evil, but can signify disaster or calamity).[44]

Other interpreters appeal to the use of the merism "good and evil" (or "good and bad") to refer to a totality.[45] Thus, the exhortation to "do good or do evil" (Isa. 41:23) means *Do something, anything!*[46] The implication of this line of thinking would be that eating of the tree represents the attempt to grasp knowledge of all things; this can be interpreted in terms of ancient notions of technology being off-limits to humans, or possibly of mantic knowledge, or in more contemporary categories of the quest for autonomy or totalization.

However, the entire phrase "knowledge of (*or* knowing) good and evil" is used in the Old Testament to refer to the normal human ability to discriminate between good and bad/evil, including the ability to make ethical decisions. Since knowledge of good and evil is precisely what Solomon asked for instead of riches (1 Kings 3:9), some have thought that the king desired what was off-limits in Genesis 2–3.[47] Yet elsewhere in the Old Testament *knowing good and evil* is taken as the legitimate ability to distinguish right from wrong, which characterizes mature adults (Deut. 1:39; Isa. 7:15), and in one case refers to the ability to discriminate with the senses, which has diminished in old age (2 Sam. 19:35 [Masoretic Text 19:36]).[48] This usage sug-

---

44. John F. A. Sawyer, "The Image of God, the Wisdom of Serpents and the Knowledge of Good and Evil," in *A Walk in the Garden: Biblical, Iconographical and Literary Images of Eden*, ed. Paul Morris and Deborah Sawyer, JSOTSup 136 (Sheffield: Sheffield Academic, 1992), 64–73.

45. A merism or *merismus* is the use of two extremes to signify not only the extremes but also everything in between.

46. In this particular text it is not the (substantive) adjectives "good" and "evil/bad," but the verbs "do good" and "do evil/harm." But the point is the same.

47. J. Daniel Hays, "Has the Narrator Come to Praise Solomon or to Bury Him? Narrative Subtlety in 1 Kings 1–11," *Journal for the Study of the Old Testament* 28, no. 2 (2003): 149–74.

48. The phrase "who today do not yet know right from wrong [lit. good and evil]" is missing from Deut. 1:39 in the Samaritan Pentateuch (but is present in the Masoretic Text and in 4QDeut^h).

gests that the tree of the knowledge of good and evil represents a normative and valuable human trait.

Since the tree seems to represent an important dimension of human maturation, some interpreters argue that Genesis 3 narrates the transgression of the divine prohibition necessary for the development of ethical decision-making. This is another form of a "fall" upwards or forwards, into maturity, becoming like God in moving beyond simple obedience to making independent ethical decisions.[49] But this interpretation is not a necessary inference.

Given the clear sense in the narrative that eating of the tree led to tragic consequences, it is better to take the tree as representing what was temporarily prohibited (for good reason), yet was not perpetually off-limits to humanity. It did not represent a form of knowledge that was reserved only for God; rather, the prohibition was dependent on timing.[50]

In accordance with what we know of moral development, children (and, by analogy, the first humans) would initially need to trust their (divine) parent, obeying parental directions for what makes for flourishing (and what to avoid), thus learning a pattern of virtue, and being formed into the sort of persons who can then (at a later stage) be allowed to eat from the tree of the knowledge of good and evil (read: decide for themselves).[51] There does, indeed, come a time in the moral development of adolescents when they need to begin making their own decisions (including ethical decisions); this is essential to the maturation process. But it makes no moral sense to allow or foster such decision-making in those without any formative experience of what is good. Eating from the prohibited tree too early would be destructive to the person, searing the conscience of the newly formed humans (we don't allow young children to "choose" between the good and evil of sexual expression and abstinence or between temperance and alcoholism or drug use). Indeed, it would both corrupt the person and lead to the violation of others (which is what happens in the Genesis account).

49. Jason P. Roberts thus thinks that humans "emerged as fallen creatures who were originally sinful." See "Emerging in the Image of God to Know Good and Evil," *Zygon* 46, no. 2 (2011): 478 (entire article 471–81).

50. This is perhaps the most significant change in my own interpretation of the garden story, since I used to think the tree represented the boundaries of finitude, beyond which it was not appropriate for humans to venture.

51. The interpretation of the tree of the knowledge of good and evil as only temporarily off-limits is a minority view in the history of interpretation, but was supported by C. S. Lewis, among others; it is central to Lewis's retelling of the garden story in his science fiction novel *Perelandra* (London: The Bodley Head, 1943).

Are there implications here for thinking about how sin began among *Homo sapiens*? Did the first humans who began to develop moral and religious consciousness go against the initial prodding of conscience and a primitive *sensus divinitatus*, and thus begin the "fall" into sin? This interpretation would have no problem with thinking that violent behavior was intermittent among (or even characteristic of) *Homo sapiens* prior to the rise of moral and religious consciousness. But such behavior becomes accountable as sin only when it is proscribed by conscience and the proscription is ignored among creatures capable of understanding the hortatory *No!*[52]

Narrative Time between Creation and Fall

Is it significant that there is no narration in Genesis 2–3 of humans fulfilling their vocation of caring for the garden? True, the *'ādām* names the animals (which partially fulfills the mandate of Gen. 1:26–28), but this is prior to the creation of the woman; she was to be a "helper" to the man, which presumably meant sharing in the task of working and protecting the garden. But instead of portraying the first humans fulfilling their explicit *raison d'être*, tending the garden (Gen. 2:15), the Genesis narrative rushes to tell of their disobedience.

Of course, the archeological record suggests that *Homo sapiens* were engaged in ordinary cultural activities such as hunting and gathering, tool-making, etc. for thousands of years prior to any evidence of the rise of moral and religious consciousness (and thus sin), which does not quite fit the narrative of Genesis.[53] In fact, the initial task given to humans in the garden story is agriculture, which bypasses the entire hunter-gatherer stage of human development. So we should not think of a strict correlation of the biblical text and evolutionary history; that would be anachronistic. Nevertheless, could the almost immediate transition from the creation of humans in Genesis 2 to the transgression in Genesis 3 be significant for thinking about the possibly limited time frame between the rise of moral and religious consciousness in *Homo*

52. It would be important to engage Paul Bloom's argument in *Just Babies: The Origins of Good and Evil* (New York: Crown; London: The Bodley Head, 2013) that even newborn infants seem to come hard-wired with a basic, though primitive, sense of morality (which includes empathy, compassion, and a sense of fairness, along with a perception of the world in terms of *us* versus *them*); this hard-wiring provides the basis for moral nurture and the development of character.

53. Genesis 4 narrates the invention of various cultural practices, including metal tools, *after* the incursion of sin.

sapiens and the onset of sin? This does not mean that the author of Genesis 2–3 knew anything of hominin evolution, but merely that the text does not actually envision a paradisiacal period. Such a period is more a function of Christian theological assumptions read back into the text than anything clearly narrated.

The Snake

The function of the snake has always puzzled thoughtful interpreters. Although the snake is identified with the devil or Satan in later tradition, in the text the snake is said to be one of the wild animals that YHWH God had made (Gen. 3:1); it is therefore (by implication) one of the animals that the human named (Gen. 2:19).[54] This point is sometimes obscured since many translations render the identical phrase *ḥayyat haśśādê* differently in Genesis 2:19 and 3:1 (the New Revised Standard Version has "animal of the field" and "wild animal," respectively).[55] Yet the point of the story is to portray the snake as a member of the (untamed) animal kingdom toward whom the human had exercised some sort of discernment, and even dominion (which seems implied by the act of naming).

That the snake is not understood as intrinsically evil is further suggested by the adjective used to describe it in Genesis 3:1. This introduction to the snake tells us that it was more "crafty" (New Revised Standard Version) than any of the other wild animals that YHWH God had made. But we must be careful to understand the meaning of this word *'ārûm*, which is translated variously as "crafty" (New Revised Standard Version, New American Standard Bible, New International Version, English Standard Version), "subtle" (King James Version), "cunning" (New King James Version, Good News Translation, Holman Christian Standard Bible), "shrewd" (New English Translation, New Living Translation), and "intelligent" (Common English Bible). This last, more neutral, rendering is important, since it indicates that the word is sometimes used as a term of approbation, to describe a wise

---

54. This is why I have intentionally used the ordinary word "snake" and not the more mythic term "serpent," as is typical in translations of Genesis 3.

55. This is probably because different translators were responsible for chap. 2 and chap. 3. Likewise, the New International Version has "all the beasts of the field" and "any of the wild animals" and the New English Translation has "every living animal of the field" and "any of the wild animals." Some translations are more consistent, such as the English Standard Version ("every beast of the field" and "any other beast of the field") and New Living Translation ("every wild animal" and "all the wild animals").

person (Prov. 12:16, 23; 13; 16; 14:8; 22:3; 27:12), where it can be rendered "prudent" (with "fool" or "simple" as its antonym).

The term does not describe what we would call a moral virtue, but more something like "street-smarts." Saul describes David's cunning with this word, since he easily escapes from him (1 Sam. 23:22). The term thus designates what we might call an *instrumental* virtue, since it names a form of intelligence that can be used for either good or evil ends. The snake is thus (initially) morally ambiguous; we don't know how it will use its intelligence.

It is also important to note that there is a pun or wordplay between this word used to describe the snake (Gen. 3:1) and the word for "naked" (*'ărûmîm*) that describes the man and woman just one verse earlier (Gen. 2:25).[56] The chapter division between these two verses should not confuse us about this important literary aspect of the narrative. The pun here is quite different in character from that between the words for human (*'ādām*) and ground (*'ădāmâ*). That wordplay indicated a primal ontological resonance between the two realities, echoing the aural resonance of the words. The same is true for the wordplay between the words for woman (*'iššâ*) and man (*'îš*), confirmed in the man's poem about the woman being bone of his bone and flesh of his flesh (Gen. 2:23). These two sets of puns indicate a fundamental unity-despite-difference between the realities named—suggested by two distinct words that nevertheless sound alike.

But the pun between *naked* and *prudent/crafty/intelligent* works in exactly the opposite way. Here we have the identical word (*'ārûm*) used with radically different meanings; the words are formally homonyms, yet they are semantically (almost) antonyms.[57] This jarring pun signals, on the seman-

---

56. The plural *'ărûmîm* is what would be expected when the adjective *'ārûm* is applied to more than one person. The singular *'ārûm* is used for nakedness in Job 24:7, 10; 26:6; Eccles. 5:14; Isa. 20:2–4; Amos 2:16.

57. Imagine a reader who has just read in Gen. 2:25 that the humans were naked (*'ārûm* in the plural) and not ashamed (already a strange idea, since nakedness in the Hebrew Bible is typically a negative quality, meaning that one is exposed and vulnerable). The reader then meets the same word for the snake just one verse later (3:1). Does this mean the snake was *naked*? Well, snakes don't have fur or feathers, so that's possible. But wait, the reader might think; *'ārûm* also means *smart/prudent/wise*. The realization of this meaning, together with the previous use of *'ārûm*, would be jarring at the semantic level, since a smart or prudent person would *never* go around naked and vulnerable. We can even see this in the English word "prude," which suggests someone who covers up and does not go about exposed. And the snake does not initially come out into the open and reveal its true motives, but rather exhibits a covert strategy of hiding and deception. Thus, the snake ultimately shows itself to be in no way "naked."

tic level, the deception the snake will perpetrate, and its instrumentality in mediating the first sin. This leads to the snake's identification in later Jewish and Christian theology with the devil or Satan, a figure who is absent from the text of Genesis 2–3.[58]

Yet the puzzle is that the snake—which, according to the logic of Genesis 1, would have been created "good"—serves as the foil to introduce temptation (and thus moral evil) into the garden story.[59] How can the snake both be part of the good created order and yet be the means of temptation or testing? How can the garden story hold *humans* accountable for the introduction of evil in the world and yet require an outside agent of temptation and sin? Perhaps an outside agent is needed to narrate a singularity such as the original sin; how else could we imagine or conceptualize evil arising in a world previously without evil?[60]

Given the above discussion of the snake, I am inclined to think that it represents that aspect of the created order which allows for, or mediates, human ethical choice. It could even be an external representation of some aspect of the human psyche (or the psyche in relationship to the external creation). Certainly, the psychological process of temptation, and the resulting sin, is vividly represented by the dialogue between the woman and the snake.[61]

The Process of Temptation and Sin

The snake's craftiness or intelligence is exhibited in the opening question he asks the woman: "Did God say, 'You shall not eat from any tree in the

58. Indeed, the rise of the figure of Satan as an independent evil persona is technically later than the Old Testament, although the common noun *śāṭān* (adversary or accuser) is often used of human beings, and in three places as a title, but not a proper name, for an angelic accuser (Job 1–2; 1 Chron. 21:1; Zech. 3:1–2).

59. Of course, there is the further conundrum that the snake speaks (the only other speaking animal in the Bible besides Balaam's donkey). Here its speech seems to be an aspect of its craft or intelligence.

60. Paul Ricoeur has struggled with the function of the snake in the text's narrating the origin of evil, given the text's emphasis on *human choice* as the origin of evil (which, Ricoeur notes, is unique among myths of origins); see Ricoeur, "The Adamic Myth," in his *The Symbolism of Evil* (Boston: Beacon, 1969).

61. Perhaps we could relate the process of temptation portrayed in the garden story to the phenomenology of temptation recounted in James 1:13–15 (where temptation involves being enticed by one's own desires).

garden'?" (Gen. 3:1). This question admits of no simple reply. Should the woman answer *yes* or *no*? Either answer would distort the truth, since God had given permission for eating of any tree in the garden—except for one. The question is technically unanswerable on its own terms.[62]

But the snake's craftiness is further shown in two changes we find when we compare the wording of his question with what the narrator says in Genesis 2. Whereas the narrator consistently uses the compound name YHWH *'ĕlohîm* to designate the creator, the snake speaks about *'ĕlohîm* only, and the woman follows suit in her response; the name YHWH is not used anywhere in their conversation (Gen. 3:1-5). Could this be a distancing tactic, to associate the prohibition with the divine realm in general and not specifically with YHWH, the God of the covenant? And beyond that, the narrator's reference to YHWH God *commanding* (Gen. 2:16) has been softened to God *saying* in the snake's question (Gen. 3:1); here again the woman follows the snake's lead (Gen. 3:3).

Yet the woman's answer to the snake is quite astute: "We may eat of the fruit of the trees in the garden; but God said, 'You shall not eat of the fruit of the tree that is in the middle of the garden, nor shall you touch it, or you shall die'" (Gen. 3:2-3.) She correctly distinguishes between the permission and the prohibition of Genesis 2:17. However, she adds the phrase "nor shall you touch it," something YHWH God never said. Is this the inner dialogue of conscience, first questioning God's word, then overstating the prohibition (building a fence around the law, to use a Rabbinic term)? Or is this dialogue suggestive of a prior conversation she may have had with the man? After all, God had given the prohibition to the man before the woman was created, so (in the logic of the story) she would likely have learned of the prohibition from the man. Did the man add the "fence," just to make it clear that this tree was "hands off"?[63] If so, could this conversation with the snake represent *interhuman* ethical deliberation? And could this be applicable to the origin of moral consciousness among *Homo sapiens*?

But there is further slippage in the woman's answer to the snake. Although the woman acknowledges the prohibition about eating from a particular tree in the garden, she vaguely describes it as "the tree in the middle

62. It is similar to the classic: "Have you stopped beating your wife?" The husband is guilty whether the reply is *yes* or *no*.

63. We might note that when instructions are passed on from an authority figure through a subordinate, the subordinate often embellishes the instructions or asserts more authority than needed (think of older siblings babysitting younger children, relaying their parents' instructions about keeping out of the cookie jar).

of the garden" (Gen. 3:3), when there were actually two trees in the middle of the garden (one was the tree of life, which was not prohibited). She also softens the warning YHWH God had given concerning the consequences of disobedience. The original warning was that *in the day* you eat of the forbidden tree you will *surely die*. But the woman omits reference to *in the day* (which suggested immediate consequences) and describes the consequence simply as "you will die" (omitting the Hebrew construction that indicated the certainty or seriousness of the consequence).

After the woman's reply, the snake asserts baldly: "You will not surely die" or (better) "You surely will not die" (Gen. 3:4), using the same construction YHWH God had previously used, but simply negating it.[64] This outright contradiction of the Creator's words shifts the dialogue to a new level—from questioning to assertion. This assertion may well play on the different meanings of *death* noted above (extinction of existence versus the incursion of corruption into life), another example of the snake's intelligence.

This bald statement is immediately followed by an explanation that impugns the Creator's motives: "for God knows that when you eat of it your eyes will be opened, and you will be like God, knowing good and evil" (Gen. 3:5). This is a half-truth, which functions to suggest that the Creator was stingy or self-protective in trying to prevent the humans from achieving knowledge that he had. This entire conversation serves to sow the seeds of doubt in the woman concerning God's generosity, resulting in a lack of trust in God's intentions for humanity.

The half-truth in the snake's final claim is evident in YHWH God's later corroboration, when he acknowledges that the humans have indeed achieved God-likeness (in Gen. 3:22). Yet, according to Genesis 1 humans were *created* in God's image (Gen. 1:26–27). They were *already* like God; it was not something they needed to attain. And this God-likeness was not connected to their knowing good and evil, but rather to their being granted dominion over the earth.[65] So when God affirms the truth of the snake's claim, that God-likeness has resulted from eating the forbidden fruit (Gen. 3:22), it has an ironic component. They have indeed become like God, but in an inappropriate way—which will not be good for them. And their eyes were indeed opened, with the result that they knew they were naked and

---

64. This is my translation; the New Revised Standard Version makes no distinction between the difference in phrasing between God, the woman, and the snake.

65. Even in the garden narrative, the infusion of the breath of life into humanity (Gen. 2:7) evoked the vivification of a cult image. Thus, in both Genesis 1 and 2 divine likeness *preceded* the transgression.

so they tried to cover their nakedness (Gen. 3:7). The sort of knowledge of good and evil they acquired was: naked = *bad*; covered = *good*.

Not only the overstating of the prohibition ("nor shall you touch it"), and the portrayal of God as stingy ("God knows that . . . you will be like God"), but also the distancing of the prohibition from the name YHWH—all these seem to fit the inner deliberation (or even interpersonal deliberation) appropriate to the phenomenology of temptation. And this could be applicable either to an "original" fall or to each person throughout history wrestling with the demands of conscience.

The result of the conversation is that the woman "saw" that the tree was "desirable" (*neḥmād*) to make one wise (Gen. 3:6).[66] This perception is stated along with her seeing that the tree was "good for food" and "a delight to the eyes," both of which correspond to similar descriptions of the trees in the garden given earlier by the narrator (Gen. 2:9). But nothing in the previous description matches the woman's new perception.

When God first animated the human from the dust of the ground (Gen. 2:7) the result was a "living soul," where *nepeš* (traditionally "soul") means something like "organism." But another translation possibility for *nepeš* is "appetite."[67] God placed the human being in the garden as a living appetite—an organism with an appetite for life; hence the reference to the garden as a source of food and beauty in Genesis 2:9. The same participle for "desirable" (*neḥmād*) in 3:6 was used earlier to describe the trees as "pleasant" to look at (Gen. 2:9). Human desire or appetite is thus appropriate and encouraged in God's world. I am inclined to think that the transference of desire from food to wisdom was not in itself wrong; wisdom is, after all, a good thing.[68] Just as the snake was not intrinsically evil, so the desire for

66. However, the Septuagint (LXX) renders the phrase "desirable to make one wise" as "beautiful to contemplate/observe," thus making it synonymous with the phrase that precedes it.

67. This is a central point in J. Gerald Janzen's analysis of Job's bitterness of "soul" (*nepeš*), in his brilliant study, *At the Scent of Water: The Ground of Hope in the Book of Job* (Grand Rapids: Eerdmans, 2009). On Janzen's reading, Job's suffering has led to the loss of his appetite for life (and with it hope for the future), which is restored only with YHWH's theophany in the sirocco/east wind (traditionally, "whirlwind"), which precedes (and portends) the fall rains after a hot, dry summer in the Arabian desert (during which the dialogue with the friends takes place). The very fact that YHWH personally appears to Job, the substance of what YHWH says, and the timing of YHWH's appearance (the Fall sirocco typically brings the scent of rain to come) all conspire to reawaken Job's *nepeš* and thus his lust for life and hope for the future—despite the terrible suffering he has experienced.

68. Indeed, the verb used in Gen. 3:6 for becoming wise (the Hiphil of *śākal*) is used of

wisdom was not in itself wrong; it was simply not the appropriate time for this momentous step.

Nevertheless, both the woman and the man take this step, with dire consequences. Yielding to her desire, the woman ate, and gave some to the man (who was there all along, but had said nothing); and he also ate (Gen. 3:6). *Eating* is here a powerful metaphor for taking something into oneself; ingesting is a participatory mode of existence, which involves making something external a part of oneself.

The Immediate Existential Consequences of the Sin

The result of this eating is an immediate existential change in the man and woman.[69] They become aware of their nakedness and—in contrast to their previous lack of shame (2:25)—they make clothing to cover themselves (3:7). Is this shame at having committed sin? Does it also represent their distrust of each other? Given that this is shared violation of a boundary God instituted (both ate of the tree), each may be wondering if the other would respect their own personal boundaries. Thus nakedness (with its implied vulnerability) is no longer safe; and from here on in the Bible nakedness is not portrayed positively.

Beyond this immediate sense of shame, the text reports their newfound fear of God, evident in their hiding when they hear God walking in the garden (Gen. 3:8). Note the answer the man gives to God's question: "I heard the sound of you in the garden, and I was afraid, because I was naked; and I hid myself" (Gen. 3:10). So even prior to the formal passing of judgment, the transgression generates (via nakedness, with its vulnerability) both shame and fear, which distances the transgressor not only from others, but also from God.

When God questions the man about whether he ate from the prohibited tree (Gen. 3:11), he blames the woman "whom you gave to be with me" (Gen. 3:12), who in turn blames the snake for deceiving her (Gen. 3:13). This refusal to take blame for one's actions (passing the buck) is a further aspect of the phenomenology of sin that reads true to life in the fallen world we know. And this finger pointing generates God's formal

---

the ultimate victory of the suffering servant in Isa. 52:13 (with the translations varying between the servant *acting wisely* and *prospering*); wisdom is meant to lead to a successful life.

69. This fulfills God's warning about "in the day" they eat of it.

declaration of judgment on the snake, the woman, and the man—in reverse order of those blamed.

The Formal Declaration of Judgment

This narration of God's declaration of judgment takes the form of a series of proclamations in poetic form (Gen. 3:14-19), which describe the consequences of the transgression, beyond the immediate existential changes that were generated.

First, the snake is redescribed using language that ironically parallels the prior statement of its cunning or intelligence (Gen. 3:14). Whereas we had been told that the snake was *cunning* (*'ārûm*) beyond all the wild animals, now it is *cursed* (*'ārûr*) beyond all the livestock and wild animals. This new pun or wordplay signifies the transformation of what was merely a creature into something negative; or perhaps it is the *relationship* between the snake and humans that is transformed. Might this portend the beginning of the process of idolatry, whereby some good aspect of God's world that has become a focus for (or mediation of) human sin is now experienced as alienating? An idol, after all, is something in creation that has become absolutized, and thereby begins to take on a negative, quasi-independent force in human affairs.[70]

The curse is then explained in terms of perpetual enmity between the offspring of the snake and the offspring of the woman (Gen. 3:15). Although an argument has been made for taking this as a protoevangelium, ever since Irenaeus (aided and abetted by the Vulgate's translation), the text clearly suggests an ongoing struggle of some sort—perhaps between humans and snakes. More likely, between humans and whatever aspects of creation become idolatrous? Perhaps even between humans and the demonic. Indeed, it is possible that this "curse" narrates the *transformation* of some aspect of creation precisely into the demonic.[71]

After the judgment on the snake comes the proclamation of judgment on the woman and then the man. These proclamations are not techni-

70. Note the paradox that Paul denies that idols exist (1 Cor. 8:4) and questions whether an idol is anything (1 Cor. 10:19), yet goes on to suggest than idols represent demons (1 Cor. 10:20).

71. For the possibility of reading the demonic as an outgrowth of the snake in Genesis 3, see Nicholas John Ansell, "The Call of Wisdom/The Voice of the Serpent: A Canonical Approach to the Tree of Knowledge," *Christian Scholars Review* 31, no. 1 (2001): 31-57.

cally *punishments*, but rather the *consequences* of human evil. Nor are they normative; they do not *prescribe* what must be. Rather, the judgments *describe* generalized consequences that men and women typically experience. These consequences not only admit of exceptions, but they are culturally conditioned, describing what is typical in the ancient social order that Israel was part of. Although the judgments that God proclaims have often been thought of as a series of "curses," neither the man nor woman is technically "cursed"—that word is used only of the snake and the ground in Genesis 3.

The typical consequences for the woman are twofold (Gen. 3:16). First, there will be an increase of pain in childbirth; that this is an *increase* of pain and not pain's origin suggests that pain is a normal response of living organisms (it does not originate with sin). And second, the man will rule the woman despite her desire for him; in other words, her yearning for intimacy will not be reciprocated. The original mutuality between the woman and the man (signified by the wordplay between 'iššâ and 'iš) will now be replaced by an asymmetry of power between them; primal resonance has become dissonance.[72]

When the narrative resumes (after the proclamation of judgment), the first thing the man does is to name the woman, thus exhibiting his rule over her; he names her Eve "because she was the mother of all living" (Gen. 3:20). Although the wordplay between "Eve" (ḥawwâ) and "living" (ḥay) suggests something beautiful and even tender, this initially positive point is contradicted by the fact of naming, which enacts an asymmetry of power.

We name animals (pets), some inanimate objects (boats), and newborn children. But once our children are grown into adults and become equal to us in status, we no longer have the authority to change their names at our whim. An example of illegitimate naming (where naming clearly enacts subjugation) is the common practice of oppressors renaming enslaved and colonized peoples. To better understand the illegitimacy of the man's naming the woman in Genesis 3:20 we should note the parallel with the naming of the animals. "The man gave names to all cattle, and to the birds of the air, and to every animal of the field; but for the man there was not found a helper as his partner" (Gen. 2:20). Since naming expresses an asymmetry of power (and humans are meant to have dominion over animals), the fact that the

---

72. Here it is important to contrast this state of affairs with the mutuality of rule granted to both male and female in Gen. 1:26-28, and also to note that the only divinely authorized rule was human rule over the nonhuman.

man named the animals showed that they did not qualify as the appropriate "helper" for him that God had intended.

Earlier God said: "It is not good that the man should be alone; I will make him a helper as his partner" (Gen. 2:18). As is well known to students of Hebrew, the term "help" or "helper" (*'ēzer* in this case; but often the participle *'ōzēr*) is typically used in the Old Testament for someone with superior power or status who comes to the aid of an inferior (Ps. 22:11 [Masoretic Text 22:12]; 72:12; 107:12; Isa. 31:3; 63:5; Jer. 47:7; Lam. 1:7; Dan. 11:34, 45]); thus God is regarded as the helper (= savior) of Israel (see Ps. 30:10 [Masoretic Text 30:11]; 54:5). But in Genesis 2:18 and 20 the word "helper" is immediately followed by *kĕnegdô*, a compound word meaning "as his partner" (New Revised Standard Version). This word qualifies "helper" so it will not be taken as a superior helper, but rather in this particular case as an equal. God intends an equality of power between the man and the woman. But naming precludes equality.[73]

It might be objected that the man previously (before the sin) had already named the woman (Gen. 2:23). Here we need to distinguish between the man's recognition of the newly created person as "woman" (*'iššâ*) taken out of "man" (*'îš*) and naming proper. The main indicator that the man does not name the woman in Genesis 2:23 is the deviation of this text from the common pattern of naming in the narratives of Genesis.[74]

In Genesis naming is typically indicated by the use of the verb *qārā'* (to call) and the noun *šēm* (name); thus Genesis 3:20 literally says, "the man called his woman's name Eve." But Genesis 2:23 uses *qārā'* (to call) without *šēm* (name); this departure from the typical pattern of naming suggests that something else is going on (plus, her name will be Eve, not "woman").[75]

---

73. Phyllis Trible is correct in her claim that the man naming the woman was an act of domination (Trible, *God and the Rhetoric of Sexuality*, 72–143). However, Trible muddies the waters by treating naming as *always* equivalent to domination. Rather, it signifies an asymmetry of power between the one doing the naming and the one being named. But some asymmetries of power are legitimate (and even nurturing), such as the relationship of parents and children.

74. George Ramsey has objected both to Trible's claim that the woman isn't named in Gen. 2:23 and to her identification of naming with domination; see Ramsey, "Is Name-Giving an Act of Domination in Genesis 2:23 and Elsewhere?" *Catholic Biblical Quarterly* 50, no. 1 (1988): 24–35. While Ramsey is mistaken in his first objection, his second objection is well taken.

75. Trible explains that although different naming formulas are used throughout the Old Testament, the Yahwist always uses the noun or verb for *name*. See Phyllis Trible, "Eve and Adam: Genesis 2–3 Reread," *Andover Newton Quarterly* 13, no. 4 (1973): 251–58.

Beyond the absence of the word for "name" in Genesis 2:23, the text uses the passive (Niphal) of *qārāʾ* ("this one shall be [or is] called woman"), which further suggests recognition of her character, rather than naming per se. The man recognizes her as one similar-yet-different from himself, indicated both by the resonant pun he makes (*ʾiššâ* taken out of *ʾîš*) and by his description of her as "bone of my bone and flesh of my flesh" (this is kinship terminology, as in 2 Sam. 5:1).

Do all women experience great pain in childbirth? Do all men dominate women? The answer to these questions is clearly *no*. These are typical human experiences in a fallen world, but they admit of exceptions. And, like all consequences of the fall (ways in which death has encroached on flourishing), they should be resisted, with remedial measures, where possible.

Following the judgment on the woman, God pronounces consequences for the man. Here the text does not use the word for man as male (*ʾîš*), but the word for human (*ʾādām*); yet *ʾādām* is treated as male (he listened to the woman; Gen. 3:17). It is curious that even after the creation of the woman the text continues to use the word *ʾādām* both for the man (Gen. 2:22, 23, 25; 3:8-9, 12, 17, 20-21) and for humanity generally (Gen 3:22-24). Does the text rhetorically enact the beginnings of patriarchy? Yet everything said of the *ʾādām* is relevant to both men and women. Because the *ʾādām* disobeyed God's word, the *ʾădāmâ* is cursed. The normative relationship of human and ground has been disrupted; primal resonance has become dissonance. This is explained in terms of the transformation of what was earlier described as "work" (*ʿābad*) into "toil" (*ʿiṣābôn*). This latter Hebrew word was already used for the woman's "pain" in childbearing. The King James Version is more democratic in translating both as "sorrow."[76]

Life and Death outside the Garden

The final consequence of the transgression is that God exiles the humans from the garden. Whereas the *ʾādām* was originally created to work (*ʿābad*) and guard (*šāmar*) the garden (Gen. 2:15), the human role is now limited to working (*ʿābad*) the ground outside the garden (Gen. 3:23). This is a significant diminution of the original human task—a task that was never actually fulfilled in the narrative of Genesis 2–3. Beyond that, it is tragic that God

---

76. And here we might note that it is not only humans who suffer after the transgression. Because the human heart has become evil (Gen. 6:5) God is also "grieved" to his heart (Gen. 6:6); the verb here is *yāṣab*, from which the noun *ʿiṣābôn* ("pain" or "toil") is derived.

has to station cherubim with a flaming sword to guard (*šāmar*) the garden—specifically the tree of life—*from* humans (Gen. 3:24), who were its original guardians.

The reason God exiles humans from the garden is to prevent them eating from the tree of life and thus living forever (Gen. 3:22). Just as God will later scatter the builders of Babel (Gen. 11:1–9), which serves to break up an imperial civilization and prevent the further concentration of power used for oppression, so here God does not want to allow the sinful human state to become permanent.[77] That this is not simple punishment, but rather a remedial act of grace, is suggested by the fact that just prior to this God clothed the naked humans with skins, something that required animal death (Gen. 3:21). And God accompanies the exiled humans outside the garden, conversing with Cain and even putting a mark of protection on him (Gen. 4:9–15).

While life outside the garden is clearly difficult (the human-earth *relationship* has been somehow disrupted), the text does not say that "nature" was changed because of the fall. It is significant that YHWH's speeches in the book of Job celebrate the wildness of the natural order, including animal predation, as glorious examples of the Creator's design of the cosmos. Many of the church fathers also celebrated natural disasters and animal predation as part of the glory of God's world.[78] So the realism of the "thorns and thistles" outside the garden fits well with what we know of the world in its natural state.

The narrative of life outside the garden in Genesis 4 may also be congruent with what we know of hominin and human evolution. Obvious connections are suggested by the often-asked questions of who Cain married (Gen. 4:17); were there other humans (or hominins) around? Perhaps the humans who were called to bear God's image with a vocation to work and protect the garden were but a representative group of *Homo sapiens*? And

---

77. The very idea of sinful immortals should remind us of the character of Q in *Star Trek: The Next Generation*. A member of the Q Continuum (a group of immortal beings), with no sense of innate morality, Q is one of the most irritating (even despicable) characters in *Star Trek*. In his immortal (and amoral) boredom, he toys with others (especially Picard) for his own amusement and intellectual stimulation.

78. Jon Garvey documents this surprising claim with abundant quotations from the church fathers, and suggests that it was not until the renaissance, when Christians began utilizing classical pagan ideas of a golden age in the distant past (from which we have declined), that the notion of a general "fall" of nature began to intrude into Christian theological writings. He discerns this shift of perspective beginning with the writings of the Reformers (Garvey, "Creation Fell in 1500" [unpublished essay]).

other questions generate similar ideas, such as why Cain took Abel into a field to kill him (Gen. 4:12) or who God was protecting Cain from when he put his mark on him (Gen. 4:15), and how many people lived in the city Cain built (Gen. 4:17).[79]

In thinking about the origin of evil, it is helpful to counterbalance the Augustinian notion of "original sin" (which assumes that all humans born thereafter come into the world enslaved to sin, by a quasi-genetic inheritance), with the actual narration of the development of sin in Genesis 4, and later in Genesis 6. The initial transgression (the "originating sin"[80]) by the parents develops in the next generation into fratricide (Cain kills Abel). But this is not a necessary progression; the narrative portrays Cain's struggle with anger and even depression (Gen. 4:5) leading up to the murder, including God's claim that he can "do well" and that although "sin is lurking at the door" he "must master it" (Gen. 4:7). God's words to Cain suggest that sin (the first use of this word in Genesis) is not an inevitability for human beings; it can (initially, at least) be resisted.[81]

The narrative of Genesis suggests a process by which humans come more and more under the sway of sin. After Cain's murder, sin grows and snowballs, evident in Lamech's revenge killing of a young man who injured him, a killing that he boasts about to his wives (Gen. 4:23), until in Genesis 6 every "inclination of the thoughts of [the human heart] was only evil continually" (Gen. 6:5), and the earth was corrupted or ruined (šāḥat) by the violence with which humans had filled it (Gen. 6:11).

Here we finally have something as pervasive as "original sin" in the later theological sense of the term—that is, a situation of communal and systemic evil we are born into.[82] The developmental aspect of how sin is portrayed in the early chapters of Genesis suggests that James is right: "sin, when it is fully grown, gives birth to death" (James 1:15). Such a developmental (and communal/systemic) view of sin as narrated in Genesis might well be

79. But perhaps this is getting too speculative (and answering these questions by appeal to other hominin groups moves too close to a new concordism).

80. This is Terence Fretheim's term for the narrative of Genesis 3; see Fretheim, *God and World in the Old Testament*, 70–76.

81. This is true even if we follow Michael Morales (with the church fathers) and translate the words for "sin is lurking/crouching at the door" in Gen. 4:7 as "a sin offering is lying at the door" (just outside the gate of the garden), and God is inviting Cain to bring a sacrifice if he fails to "do well." See L. Michael Morales, "Crouching Demon, Hidden Lamb: Resurrecting an Exegetical Fossil in Genesis 4.7," *The Bible Translator* 63, no. 4 (2012): 185–91.

82. Suggested by Fretheim, *God and World in the Old Testament*, 70, 79.

suggestive for thinking about the growth of moral evil among early *Homo sapiens*.

Quo Vadis?

I am quite aware that this is only an introductory exploration of Genesis 3 with respect to human evolution. I am under no illusions that I have any clear answers to how we should think about the biblical account of the fall together with the origin of evil among *Homo sapiens*. It is no simple matter to bring together our biblical inheritance and the realities of biological evolution in a Chalcedonian spirit, without confusing or separating the discourses (the "natures")—"the distinction of natures being by no means taken away by the union, but rather the property of each nature being preserved" (as the Chalcedonian Creed puts it). If crude concordism mingles the natures, perhaps NOMA separates them too distinctly.

At times I have wondered where my exploration of the biblical text was leading. But I judged that we needed a close reading of the text's theological motifs in order to prevent our being immediately overwhelmed by the claims of contemporary science. Perhaps indwelling our formative narratives of creation and fall, with our eyes open to what we know (or think we know) about human evolution, is an adequate first step.

5 "Adam, What Have You Done?"

New Testament Voices on the Origins of Sin

JOEL B. GREEN

Reinhold Niebuhr famously acknowledged that "the doctrine of original sin is the only empirically verifiable doctrine of the Christian faith."[1] Whether we are ready to co-sign this claim depends on what is meant by "original sin." Clearly, it is easy to counter modern optimism regarding human dignity, virtue, and happiness by drawing attention to our surplus of sinful acts, past and present. This is the argument John Wesley mounted in his treatise, *The Doctrine of Original Sin according to Scripture, Reason, and Experience* (1757)—his attempt to ground the need for everyone to repent. The doctrine of original sin has traditionally concerned itself with more than people behaving badly, however, and observations regarding sinful behavior should not be confused with diagnoses of sin's etiology. Unsurprisingly, then, a judgment like Niebuhr's can readily be associated with a chastened view of original sin, one that has less to do with origins and more to do with the human family's blindness to its involvement in sin.[2]

Indeed, traditional notions of original sin have increasingly fallen on hard times. For many, the notion that we all might be held accountable for the misdeed of a single, common ancestor or first couple boggles the mind as a historical and/or moral nonstarter. In the wake of evolutionary

---

1. Reinhold Niebuhr, *Man's Nature and His Communities: Essays on the Dynamics and Enigmas of Man's Personal and Social Existence* (New York: Charles Scribner's Sons, 1965; repr., Eugene, OR: Wipf & Stock, 2012), 24 (citing the *London Times* with approval).

2. Thus, Tatha Wiley, *Original Sin: Origins, Developments, Contemporary Meanings* (Mahwah, NJ: Paulist, 2002), 208.

biology, the idea of a single, original couple, together with the idea that human history might be divided into two eras, pre- and post-Fall, seems absurd to many. As a result, some have called for the dismantling of the doctrine altogether and others for its dogmatic reformulation;[3] more generally, perhaps, the doctrine suffers from benign neglect. It is not too much to say that the particulars of the doctrine of original sin have been and are the subject of ongoing negotiation. Original sin may belong to our Christian heritage, particularly though not exclusively among Protestants, but, absent from the church's ecumenical creeds, it lacks magisterial definition for the whole church. We might naturally ask, then, whether scripture orients us regarding sin's origins, or how scripture might inform our formulation(s) of the doctrine of original sin.

My particular interest is with the contribution of the New Testament to this doctrine. I will approach the question of sin's origins, first, by sketching briefly how Jewish texts in the Second Temple period understood the significance of Adam's sin. This will help us to frame our reading of two New Testament authors who reflect on the character of sin: Paul and James. Neither of these New Testament theologians refers to an event we might term "the Fall," but they do have something to say about the nature of sin, its universality, and its practical inevitability.

"Original Sin" in the Second Temple Period

How did Adam and Eve's disobedience fare at the hands of Jewish writers? Interestingly, beyond the opening chapters of Genesis itself, the Old Testament reflects little on Adam and Eve or on what happened in the Garden. Only twice is Adam mentioned in connection with sin, both times in the form of a simile. Job wonders whether, like Adam, he has concealed his offense (Job 31:33).[4] And speaking of God's people, Hosea writes, "Like

---

3. Dismantling—e.g., John E. Toews, *The Story of Original Sin* (Eugene, OR: Pickwick, 2013); Patricia A. Williams, *Doing without Adam and Eve: Sociobiology and Original Sin*, Theology and the Sciences (Minneapolis: Fortress, 2001). Reformulation—e.g., Wiley, *Original Sin*; Daryl P. Domning and Monika K. Hellwig, *Original Selfishness: Original Sin and Evil in the Light of Evolution* (Aldershot, UK: Ashgate, 2006); Ian A. McFarland, *In Adam's Fall: A Meditation on the Christian Doctrine of Original Sin*, Challenges in Contemporary Theology (Malden, MA: Wiley-Blackwell, 2010).

4. So, e.g., the Authorized Version and the Common English Bible; other translations render *kĕ'ādām* with references to "people" (e.g., New Revised Standard Version, New Liv-

Adam they broke the covenant" (Hos. 6:7). The ramifications of Adam's or Adam and Eve's disobedience in the Garden are not discussed, though subsequent chapters in Genesis do represent sin as though it were a human epidemic: Cain's murderous act leads to his exile (Gen. 4:1–16); a restless, godless society emerges (Gen. 4:17–24; 5:28–29); global violence leads to global destruction (Gen. 6:1–9:18); sin among Noah's family leads to one people enslaving another (Gen. 9:17–27); and, finally, humanity is implicated in an imperialist plot, giving rise to a tower made to touch the heavens (Gen. 11:1–10). Like a contagion, one sin leads to the next, until, like a virus, it has spread to the entirety of human existence. Much can be said about sin, then, even when the problem of sin's etiology is not in focus.

For early Jewish reflection on Adam's significance we turn to writings from the roughly three-hundred-year period from the early second century BCE to the late first or early second century CE.[5] Here we find some alternative viewpoints on the ramifications of Adam's sin for later generations. These are important for our inquiry because they set something of the terms of the discussion that would have been in the air Paul and James breathed. Among these texts, four are immediately relevant to our question concerning the consequences of the story told in Genesis 3 for the human family: the *Life of Adam and Eve* (also known as the *Apocalypse of Moses*; late first century CE?), *4 Ezra* (end of the first century CE), *2 Baruch* (late first or early second century CE), and *Biblical Antiquities* (also known as Pseudo-Philo; first century CE).

Life of Adam and Eve

The *Life of Adam and Eve* tells the story of what happened after Adam and Eve were expelled from the Garden of Eden, with Adam and Eve serving as representative humans in search of food, facing temptation, experiencing pain in childbirth, and so on. Late in the story, Adam, faced with death, tries to explain sickness and pain to their children. In an account reminiscent of Genesis 3,

---

ing Translation). Unless otherwise indicated, biblical citations follow the Common English Bible.

5. For surveys, see John J. Collins, "Before the Fall: The Earliest Interpretations of Adam and Eve," in *The Idea of Biblical Interpretation: Essays in Honor of James L. Kugel*, ed. Hindy Najman and Judith H. Newman, Supplements to the Journal for the Study of Judaism 83 (Leiden: Brill, 2004), 293–308; John R. Levison, "Adam and Eve," in *Eerdmans Dictionary of Early Judaism*, ed. John J. Collins and Daniel C. Harlow (Grand Rapids: Eerdmans, 2010), 300–302.

Adam recalls that God brought plagues on Adam because Adam had rejected God's covenant (*Life of Adam and Eve* 8:1). Eve bears particular responsibility in this story, though, as it is "through" her that Adam dies (*Life of Adam and Eve* 7:1).[6] She confesses to Adam that "this has happened to you through me; because of me you suffer troubles and pains" (*Life of Adam and Eve* 9:2), and later laments, "Woe is me! For when I come to the day of resurrection, all who have sinned will curse me, saying that Eve did not keep the command of God" (*Life of Adam and Eve* 10:2; cf. 14:2). After Adam's death, she exclaims,

> I have sinned, O God; I have sinned, O Father of all; I have sinned against you. I have sinned against your chosen angels, I have sinned against the cherubim, I have sinned against your steadfast throne; I have sinned, LORD, I have sinned much; I have sinned before you, and all sin in creation has come about through me. (*Life of Adam and Eve* 32:2–3)

Sin, we learn, resulted from Eve's encounter with the serpent who "sprinkled his evil poison"; this is the poison of "desire" or "craving" (*epithymia*),[7] "the origin of every sin" (*Life of Adam and Eve* 19:3).

Clearly, according to the *Life of Adam and Eve*, death originates with Adam and Eve's sin. They are (or she is) the fountainhead of sin, even if this does not remove responsibility for sinful inclinations and behaviors from future generations. Sin results from craving, the source of which is the serpent. Eve's story of how she and Adam were deceived (*Life of Adam and Eve* 15–30) has a pedagogical purpose, designed to serve as a prophylactic against future disobedience: "Now, then, my children, I have shown you the way in which we were deceived. But you watch yourselves so that you do not forsake the good" (*Life of Adam and Eve* 30:1).

4 Ezra

In *4 Ezra* we find a series of interactions between Ezra and God, or between Ezra and God's angel Uriel, concerning Israel's exilic situation.[8]

---

6. English translations of *Life of Adam and Eve* are taken from M. D. Johnson, "Life of Adam and Eve: A New Translation and Introduction," in *The Old Testament Pseudepigrapha*, 2 vols., ed. James H. Charlesworth (Garden City, NY: Doubleday, 1983), 2:249–93; my references are to the Greek version.

7. M. D. Johnson translates "covetousness" ("Life of Adam and Eve," 279).

8. *4 Ezra* comprises chaps. 3–14 of the apocryphal book *2 Esdras*, which is itself a com-

Here is the problem: given the overwhelming power of the human inclination toward evil, why has God abandoned his people? Ezra wants
to trace suffering and death to Adam's disobedience: "The first Adam,
burdened with this inclination [to do evil], disobeyed you and was overcome, but so were all those descended from him. The disease became
permanent; the Law was in the people's heart along with the wicked
root, and that which was good departed and the wickedness remained"
(4 Ezra 3:21–22).[9]

First, then, Ezra recognizes the culpability of a long line of human
beings: "each nation lived by its own will, and people acted without giving you a thought. They acted with scorn, and [the Lord] didn't prevent
them" (4 Ezra 3:8). Second, Adam himself was burdened with an evil
inclination: "A grain of evil seed was sown in the heart of Adam from the
beginning, and how much godlessness it has produced until now and will
produce until the time for threshing comes!" (4 Ezra 4:30). Why does
the heart lean toward evil (cf. 4 Ezra 4:4)? We are never told, though we
do discover that the human inclination toward doing evil can and should
be countered through the exercise of free will in the service of the Law
(e.g., 4 Ezra 7:19–24, 118–26; 8:46–62; 14:34). Ezra traces mortality back
to Adam and faults the Lord for not taking away the evil inclination, but
4 Ezra identifies neither God nor Adam as the ultimate source of this
disposition to sin.[10]

A second, related issue between Ezra and Uriel is Ezra's concern for the
fate of a sinful human family. Ezra is dismayed to learn that, at the end-time
judgment, the door will be closed on further offers of mercy. Keenly aware
of the ubiquity of sin, he responds:

> Adam, what have you done?! If you sinned, the downfall wasn't yours
> alone but also ours who are descended from you. What benefit is it to
> us that we are promised an immortal time, but we have done works
> that bring death? What good is it to us that everlasting hope has been
> predicted for us, but we have utterly failed? What good is it that safe
> and healthy dwelling places are reserved, but we have behaved badly?

pilation of three books: (1) chaps. 1–2, sometimes designated as 2 or 5 Ezra; (2) chaps. 3–14,
usually called 4 Ezra; and (3) chaps. 15–16, sometimes designated as 5 or 6 Ezra.

9. English translations of 4 Ezra follow the Common English Bible.

10. In 7:116, Ezra apparently assumes the earth's culpability: "It would have been better
if the earth hadn't brought forth Adam, or when it had brought him forth, that it had forced
him not to sin."

What good is it that the glory of the Most High will protect those who have conducted themselves decently, but we have conducted ourselves indecently? What good is it that paradise will be revealed, whose fruit remains uncorrupted, in which there is plenty and healing, but we won't enter it, for we have visited unseemly places? What good is it that the faces of those who practiced abstinence will shine brighter than stars when our faces are blacker than darkness? While we were alive and doing evil, we didn't think about what we would suffer after death. (*4 Ezra* 7:118–26)

Here is the basic dilemma: God has purposed good things for humanity, but humanity will not experience those good things on account of sin's pervasiveness.

Adam's sin marks the downfall of all of his descendants, yet Adam is not the only offender. Humans after Adam, we are told, have "done works that bring death," "utterly failed," "behaved badly," "conducted ourselves indecently," and "visited unseemly places." Leaning toward the bad, all people bear responsibility for their own actions. If only humans would follow God's instruction! "If then you will rule your mind and instruct your heart, you will be kept alive, and after death you will attain mercy. Judgment comes after death, when we are restored to life, and then the names of the just will appear and the deeds of the wicked will be exposed" (*4 Ezra* 14:34–35).

2 Baruch

2 Baruch is a revelation concerning the recent destruction of Jerusalem in 70 CE and its ramifications for Jewish life. Seeking to make sense of this national trauma, the book mentions Adam during three dialogues between Baruch and God (*2 Bar.* 13:1–20:6; 22:1–30:5; 48:26–52:7). The first two identify Adam's sin as the beginning of death, but say nothing about sin's origins or inheritability. In the third exchange, Baruch exclaims, "O Adam, what did you do to all who were born after you? And what will be said of the first Eve who obeyed the serpent, so that this whole multitude is going to corruption?" (*2 Bar.* 48:42–43). With these words Baruch does not lay responsibility for human sin and for divine judgment at Adam and Eve's feet. Instead, he goes on to speak of all who fail to recognize God as their creator and who disobey the law (*2 Bar.* 48:46–47), and to underscore human culpability:

And those who do not love your Law are justly perishing. And the torment of judgment will fall upon those who have not subjected themselves to your power. For although Adam sinned first and brought death upon all who were not in his own time, yet each of them who has been born from him has prepared for himself [*sic*] the coming torment. And further, each of them has chosen for himself the coming glory. For truly, the one who believes will receive reward. (*2 Bar.* 54:14–16)

Accordingly, although Adam brought mortality to this age, people have the capacity to determine their future destinies. People are sinners because they sin; or, as Baruch himself concludes, "Adam is, therefore, not the cause, except only for himself, but each of us has become our own Adam" (*2 Bar.* 54:19).

Biblical Antiquities

As an example of the genre of "rewritten Scripture," *Biblical Antiquities* recounts the biblical story from Adam to Saul's death. A pivotal reference to Adam appears in a retelling of God's instruction to Noah after the Flood. Speaking of "paradise," God said:

This is the place concerning which I taught the first man, saying, "If you do not transgress what I have commanded you, all things will be subject to you." But that man transgressed my ways and was persuaded by his wife; and she was deceived by the serpent. And then death was ordained for the generations of men [*sic*]. (*Bib. Ant.* 13:8)[11]

Speaking now of Moses, whom the writer of *Biblical Antiquities* regards as the author of Genesis, the text continues, "And the LORD continued to show him the ways of paradise and said to him, 'These are the ways that men [*sic*] have lost by not walking in them, because they sinned against me'" (*Bib. Ant.* 13:9). Two observations clearly follow: Adam's sin results in human mortality and God's people are themselves responsible for their own obedience (or lack of obedience). *Biblical Antiquities* demonstrates no interest

---

11. English translations of *Biblical Antiquities* are taken from D. J. Harrington, "Pseudo-Philo: A New Translation and Introduction," in *The Old Testament Pseudepigrapha*, 2 vols., ed. James H. Charlesworth (Garden City, NY: Doubleday, 1983), 2:297–377.

in the particular problem of sin's origins—a point that gains in importance on account of the scholarly consensus that this book represents mainstream Jewish interpretation of scripture in first-century Palestine.[12]

Reflections

Although Israel's scriptures are themselves bereft of theological reflection on the ongoing significance of Adam and Eve's disobedience in the Garden, a few Jewish texts from the Second Temple period do work with Genesis 3 as they tell something of the story of sin. These texts agree in two important respects: (1) Adam (or Eve's) disobedience results in their own mortality and in the mortality of all who would come after them, and (2) human beings remain responsible for their own actions.[13]

In two of these texts, sin's cause is a matter of interest. In the *Life of Adam and Eve*, the serpent gives Eve—and apparently with her, the whole human family—the toxin of "desire" or "craving," that is, a heart inclined toward evil. In *4 Ezra*, a characteristic feature of humanity is the "evil inclination," though its origins are unclear. The idea of an evil inclination is found in other Jewish texts as well,[14] though human beings are never said to be inherently under its control. Rather, humans remain free to choose the good, with the good typically identified with God's instruction.

In short, when sin's origins are discussed, Jewish writers of the Second Temple period refer to human choice even as they speak of Adam's (or Adam and Eve's) influence. Sin is not compulsory, even if its ubiquity might suggest its inevitability.

12. On this point, cf. Frederick J. Murphy, "Biblical Antiquities (Pseudo-Philo)," in *Eerdmans Dictionary of Early Judaism*, ed. Collins and Harlow, 440–42 (here, 442).

13. See the similar conclusion in Thomas H. Tobin, *Paul's Rhetoric in Its Contexts: The Argument of Romans* (Peabody, MA: Hendrickson, 2004), 171–74 (especially 172); Peter C. Bouteneff, *Beginnings: Ancient Christian Readings of the Biblical Creation Narratives* (Grand Rapids: Baker Academic, 2008), 9–26 (especially 26); Toews, *Original Sin*, 37.

14. Cf., e.g., Sirach 5:2; 15:14–15; 19:30; 23:4–5; Philo, *Special Laws* 2:163; 4 Maccabees 1:1; CD 2:15.

New Testament Voices

Paul and James would have swum in the same theological pond as the Jewish writers we have discussed thus far, and their contributions can be read in dialogue with perspectives like those of *4 Ezra* or *Biblical Antiquities*.

Paul and Sin's Power

Contemporary readers of the New Testament might be tempted to think of sin primarily in autobiographical and individualistic ways, finding support for such views in phrases like "the forgiveness of sins." Even though, in first-century usage, this phrase is remarkably corporate in its reach, referring as it does especially to God's restoration of God's people, it is worth noting that the phrase "forgiveness of sins" is surprisingly rare in letters attributed to Paul, appearing only in Ephesians 1:7 and Colossians 1:14. This is because of the way Paul understands "sin." To anticipate what follows, for Paul "sin" is more a power from which humans need to be liberated than individual, wrongful deeds for which humans require forgiveness.

Among the letters written by or attributed to Paul in the New Testament, *hamartia* ("sin") appears sixty-four times, congregating above all in his letter to the Romans.[15] Of the thirty-nine appearances of *hamartia* in Romans, thirty are found in Romans 5–7. Forms of the verb *harmartanō* ("I sin") appear in the Pauline corpus fourteen times—six in Romans and four in Romans 5–7. The related adjectival form, *hamartōlos* ("sinful," "sinner"), appears eight times in the Pauline letters, four times in Romans, and three in Romans 5–7. Although Paul's vocabulary of sin is not limited to the appearance of this one word-group, the accumulation of the language of "sin" in these three chapters of Romans nevertheless calls attention to the importance of this section of the letter for understanding Paul's perspective on sin.

Especially remarkable is the agency allotted to sin. Sin "entered the world" (Rom. 5:12), where it exercised power reminiscent of a master-slave or king-subject relationship. Humans are "controlled by sin" (Rom. 6:6), the aim of sin is to "rule your bodies, to make you obey their cravings [*epi-*

---

15. I have adapted some of this material on Paul from Joel B. Green, *Body, Soul, and Human Life: The Nature of Humanity in the Bible*, Studies in Theological Interpretation (Grand Rapids: Baker Academic, 2008), 98–100.

thymia]" (Rom. 6:12, my translation), people present "parts of their body to sin as weapons to do wrong" (Rom. 6:13), and people comport themselves as slaves to sin (Rom. 6:16). Whereas in the *Life of Adam and Eve*, cravings give rise to sin, for Paul the opposite is the case: sin produces all kinds of cravings. And whereas in the Second Temple Jewish literature we surveyed obedience to the law was the means for countering cravings and sin, for Paul the law is an instrument sin uses to cultivate those cravings (*epithymia*; Rom. 6:12; 7:7–8). The baptized were formerly enslaved to sin, but they are now liberated from its dominion (Rom. 6:17–18, 20, 22).

We might summarize Romans 6, then, with reference to the inevitability of human servitude, with the only question being the identity of the master to whom one's life is presented: "to sin, to be used as weapons to do wrong," or "to God as weapons to do right" (v. 13). Paul reasons, "Once, you offered the parts of your body to be used as slaves to impurity and to lawless behavior that leads to still more lawless behavior. Now, you should present the parts of your body as slaves to righteousness, which makes your lives holy" (v. 19). Again: "Don't you know that if you offer yourselves to someone as obedient slaves, that you are slaves of the one whom you obey? That's true whether you serve as slaves of sin, which leads to death, or as slaves of the kind of obedience that leads to righteousness" (v. 16). Thus for Paul, baptism is union with Christ's death and rising to new life, with the result that those who have undergone baptism no longer bow to sin's hegemony.

Among the texts to which we might turn to see how Paul borrows from the story of Adam as he paints the human situation, two in his letter to the Romans are of special interest. The first is Romans 1:18–32, the opening of an argument leading to the conclusion that "both Jews and Greeks are all under the power of sin" (Rom. 3:9); stated differently, "all have sinned and fall short of God's glory" (Rom. 3:23). In Paul's diagnosis, "sins," identified as human impiety and wrongdoing, arise from "sin," a general disposition to refuse to honor God as God and to render him thanks. In Romans 1:18–32, Paul is not concerned with providing the biography of individuals, as though he were demarcating the steps by which a person falls into sin or comes to be implicated in sin. Rather, his is a universalistic presentation, an analysis of the human situation understood corporately, as God hands the human family over to experience the consequences of the sin they choose (Rom. 1:18, 24, 26, 28). James Dunn has recognized that Adam stands in the background of Paul's argument. Linguistic hints connecting Romans 1 and Genesis 2–3 are minimal, but any possible echoes (for example, the heightened emphasis on "knowing"[16]) are strengthened by the opening words of Romans 1:20: "Ever since the creation of the world. . . ." In effect, the story of the entire human family who chose creation over the creator only to be given over to their own desires and distortions—this cosmic story reflects the story of the one human, Adam. In Dunn's words, Romans 1:18–32 reflects the life of Adam "who perverted his knowledge of God and sought to escape the status of creature," thus setting the pattern for the idolatry that would characterize Israel and, indeed, all humanity.[17]

The second text is Romans 5:12–21, where Adam is no longer lurking in the shadows of Paul's rhetoric but is explicitly named, so that Paul can contrast the deeds of Adam and of Jesus Christ. Earlier in Romans 5, Paul repeatedly uses the first-person plural pronoun "we," unmistakably referring to Christ-followers: "we have been made righteous," "we have peace with God," and so on (Rom. 5:1–11). With the conspicuous shift to the third person in Romans 5:12, Paul now turns to map the world of humanity writ large. Bracketing this section of the letter, then, are phrases that document the changed situation for humanity: "Just as through one human being sin came into the world, and death came through sin . . . so also grace will rule through righteousness to bring eternal life through Jesus Christ our Lord" (Rom. 5:12, 21; my translation).[18]

Romans 5:12 is the pivotal text, since here we have what might be regarded as Paul's reflection on the effects of Genesis 3: "Just as through one human being sin came into the world, and death came through sin, so death has come to everyone, since everyone has sinned" (my translation). Of particular interest is the last phrase of the verse, *eph' hō pantes hēmarton*, which I have translated "since everyone has sinned." The fourth-century commentator Ambrosiaster (followed by Augustine) took this phrase as a reference to humanity's having sinned "in Adam," but contemporary translations regard it as a marker of cause: "because all [have] sinned."[19] In actuality, the debate on the sense of this phrase has long quieted in favor

---

16. Cf. the repetition of *gnōstos* ("known") in Gen. 2:9 and Rom. 1:19, and the use of the verbal form *ginōskō* ("I know") in Gen. 2:17; 3:5, 7, 22 and Rom. 1:21.

17. James D. G. Dunn, *Romans 1–8*, Word Biblical Commentary 38A (Dallas: Word, 1988), 53.

18. Following Martinus C. de Boer, "Paul's Mythologizing Program in Romans 5–8," in *Apocalyptic Paul: Cosmos and Anthropos in Romans 5–8*, ed. Beverly Roberts Gaventa (Waco, TX: Baylor University Press, 2013), 1–20 (here, 8).

19. E.g., New English Translation, New International Version (2011), New Revised Standard Version; New Jerusalem Bible: "because everyone has sinned."

of this latter reading.[20] Paul thus grounds universal human mortality in his affirmation of the universality of sin, an affirmation he can make on the basis of a phenomenological observation—"because all sinned"—which he interprets theologically in relation to Adam's sin.

How this is so merits careful study. On the one hand, *through* Adam and *from* Adam, sin entered the world, death ruled, many people died, judgment came, many people were made sinners, and sin ruled in death (Rom. 5:12–21). On the other hand, death came to everyone because everyone sinned. Indeed, Paul had affirmed earlier in the letter that "all have sinned and fall short of God's glory" (Rom. 3:23), and "God will repay everyone based on their works" (Rom. 2:6): "wrath and anger for those who obey wickedness" but "glory, honor and peace for everyone who does what is good" (Rom. 2:10). We might say, then, both that Sin (with an uppercase "s," a malevolent power) entered the world on account of Adam, and that Adam's disobedience set in motion a chain of effects, one sin leading to the next, not because sin was an essential constituent of the human condition but because Adam served as the pattern all humanity followed in his sinfulness.

Actually, the larger thesis Paul is advancing has less to do with Adam and more to do with Christ. He wants to show that Christ is the savior of all humanity, both Jew and Gentile. What he needs to demonstrate, then, is that Jew and Gentile stand on level ground with respect to their need for salvation. What Paul requires, then, is this affirmation of human solidarity, Jew and Gentile, in sin—and, therefore, in death. The first part of this affirmation is in place already, having been set out in Romans 1–3. Accordingly, Paul's appeal to Adam's sin in Romans 5 plays a supportive or adjunctive role to that earlier material. Paul can assume in Romans 5 the universality of sin as a warrant for his further affirmation. This is that all humanity is implicated in death, so that all stand in need of "eternal life through Jesus Christ our Lord" (Rom. 5:21).

---

20. Among the grammarians, cf., e.g., J. H. Moulton, W. F. Howard, and N. Turner, *A Grammar of New Testament Greek*, vol. 1 (Edinburgh: T. & T. Clark, 1908), 107: "in view of the fact that"; F. Blass, A. Debrunner, and R. W. Funk, *A Greek Grammar of the New Testament and Other Early Christian Literature* (Chicago: University of Chicago Press, 1961), §235(2): "for the reason that, because"; C. F. D. Moule, *An Idiom Book of New Testament Greek*, 2nd ed. (Cambridge: Cambridge University Press, 1959), 50: "inasmuch as, because"; Daniel B. Wallace, *Greek Grammar beyond the Basics: An Exegetical Syntax of the New Testament* (Grand Rapids: Zondervan, 1996), 342–43: "because." Among commentators, see especially C. E. B. Cranfield, *A Critical and Exegetical Commentary on the Epistle to the Romans*, International Critical Commentary, vol. 1 (Edinburgh: T. & T. Clark, 1975), 274–81.

Although we have examined only a portion of Paul's work, what we have studied is nonetheless important for its concern with the ongoing effects of Adam's disobedience as Genesis 3 tells the story. With regard to sin, Paul underscores the solidarity of the entire human family in sin, not because Adam's sin was or is somehow imputed to the human race but because everyone follows Adam in sinning. Furthermore, Adam's disobedience introduces Sin as a hegemonic force in the world. Paradoxically, therefore, human sinfulness is for the apostle a sign of both human helplessness and culpability. As he will go on to affirm in Romans 7, sin is an active agent at work in human transgression, performing like an alien intruder at work in the sinner (vv. 17, 20). As a result, for Paul, the potency of sin as the author of human behavior is not a manifestation of human perversity, but of human frailty.[21] Paul's diagnosis is thus significantly more radical than what we find in the Second Temple Jewish literature we surveyed earlier. Adam's sin sets the pattern for human sin more broadly, to be sure. But for Paul sin itself is a power that cannot be overcome through obedience to God's instruction, not because God's instruction is defective but because human efforts are inadequate in the face of sin's pull. Accordingly, all humanity stands in need of the life available in Christ.

James on Sin's Origins

The letter of James provides a keen diagnosis of the problem of sin.[22] One window into James's thought is provided in 4:4: "Adulterers! Don't you know that friendship with the world means hostility toward God? Therefore, whoever wants to be the world's friend becomes God's enemy" (my translation). For Luke Timothy Johnson, this verse is the theological center of the letter's moral exhortation.[23] James's metaphor depends on classical ideas of friendship, understood in terms of unity of heart and mind. Cicero

21. Cf. Udo Schnelle, *The Human Condition: Anthropology in the Teachings of Jesus, Paul, and John* (Minneapolis: Fortress, 1996), 63–66; Klaus Berger, *Identity and Experience in the New Testament* (Minneapolis: Fortress, 2003), 207–9.

22. I have adapted some of this material on James from Green, *Body, Soul, and Human Life*, 94–98.

23. Luke Timothy Johnson, *The Letter of James: A New Translation with Introduction and Commentary*, Anchor Bible 37A (New York: Doubleday, 1995), 80–88; "Friendship with the World and Friendship with God: A Study of Discipleship in James," in *Brother of Jesus, Friend of God: Studies in the Letter of James* (Grand Rapids: Eerdmans, 2004), 202–20.

(first century BCE), for example, described friendship as "nothing other than the agreement over all things divine and human along with good will and affection" (*On Friendship* 6.20).[24] Given his unrelentingly negative portrait of the "world" (James 1:27; 2:5; 3:6; 4:4), James's polarized rhetoric, pitting friendship with the world against friendship with God, hardly seems surprising. And with the frankness of the choices confronting Christ-followers, we may not be surprised to find James's invective unleashed against the "double-minded" (*dipsychos*), another word that helps us to understand the character of sin for James (James 1:7; 4:8; cf. Ps. 119:113). Lacking a pure heart, the double-minded deceive themselves—thinking they are truly devoted to God when their affections and behavior broadcast the opposite. Finally, when James labels his audience as "adulterers" (James 4:4), he echoes the biblical tradition of Israel as God's unfaithful wife, that is, as those having (or claiming to have) a covenant relationship with Yahweh while engaging in idolatry.

Insight into what comprises "friendship with the world" can be found in a number of texts in this letter, but perhaps none is more transparent than its depiction of "earthly" wisdom in terms of "jealousy and selfishness," which gives rise to "disorder and every evil practice" (James 3:14-16). This is the converse of the wisdom God gives, which "is first pure, then peaceable, gentle, accommodating, full of mercy and good fruit, impartial, and not hypocritical" (James 3:17; see 1:5). We easily see, then, that for James the epitome of the sinful life is not an act but an allegiance, relationally delimited: "friendship with the world."

Another avenue into James's perspective on sin, and one more promising for our agenda, appears in the letter's first chapter, where he speaks to his audience's experience of *peirasmos* ("trial" or "temptation"). First, James sketches a progression from *trials (peirasmos)* to maturity:

> My brothers and sisters, consider it nothing but joy when you fall into all sorts of trials, because you know that the testing of your faith produces endurance. And let endurance have its perfect effect, so that you will be perfect and complete, not deficient in anything. (James 1:2-4)

Then, he sketches a parallel progression from trials, now understood as *temptations (peirasmos)*, to death:

24. See, e.g., David Konstan, *Friendship in the Classical World* (Cambridge: Cambridge University Press, 1997).

> Everyone is tempted by their own cravings; they are lured away and en-
> ticed by them. Once those cravings conceive, they give birth to sin; and
> when sin grows up, it gives birth to death. (James 1:14–15)

Between these two chains of effects lies a pronouncement of blessing for all
who endure *peirasmos*, as well as James's reply to the all-important question
he anticipates: What is the source of *peirasmos*?

James's argument at this juncture invites special attention because James
1:13–18 can be read as his theological reflection on Genesis 1–3 in order to
buttress his claim regarding God's character and gifts. Sharply put, God is
not the source of temptation:

> Don't be misled, my dear brothers and sisters. Every good gift, every per-
> fect gift, comes from above. These gifts come down from the Father, the
> creator of the heavenly lights, in whose character there is no change at all.
> He chose to give us birth by his true word, and here is the result: we are like
> the first crop from the harvest of everything he created. (James 1:16–18)

The Common English Bible, cited above, helpfully clarifies that James uses
the phrase "Father of lights" to refer to God as creator and, thus, to creation
(see Gen. 1:3, 14–17). This God does not send temptation (James 1:13–15), but
good things. This God does not waver between giving good and bad things,
for his character is consistently oriented toward providing good things. And
this God is known for his generosity. He gives "without a second thought,
without keeping score" (James 1:5), and he gives "every good gift, every
perfect gift" (James 1:17).

James emphatically denies that *peirasmos* originates with God (James
1:13), but instead traces *peirasmos* to desire, or craving (*epithymia*). Here we
find James's dexterity with the term *peirasmos* in all its semantic ambiguity.
In the biblical tradition, it can refer to *diabolic temptation* (which impedes
and corrupts human life) just as easily as to *divine test* (which refines and
deepens human life). Recall that James addresses his letter to exilic sojourn-
ers (James 1:1) who, like exiles more generally, face distress and conflict;
should these be characterized as testing or as temptation? What determines
for James whether their *peirasmos* takes the form of "testing" or of "temptation" is not the experience itself but *human desire* or *human craving*, that
is, whether people master their sinful dispositions or are mastered by them.
This does not mean that James regards the human person as inherently evil,
even if we should recognize that, for James, humans are characterized by

an inclination toward evil even as they continue to bear the image of God in which they were made (James 3:9).[25]

James associates "desire" with two semantic fields, both viewed as an almost irrepressible force: a fisherman with his favorite fishing lure and an irresistible seductress (James 1:14).[26] With his reference to "desire," then, James identifies the source of his audience's real difficulties not in terms of external pressures and certainly not as manifestations of the divine will, but as internal inclinations. *Epithymia*, used in James 1:14–15,[27] can be taken in the neutral sense of "desire," but in moral discourse typically the term has the negative sense of "evil desire" or "craving." Here, its generative role with respect to sin and death unmistakably qualifies it as negative and associates it with the wider Jewish tradition of the evil inclination (see above), and it is this evil inclination that gives rise to the double-mindedness by which James characterizes sin.[28] Personal responsibility is further underscored by the addition of *idios* ("one's own") in James 1:14: "his or her own craving" (my translation).

In the verses that follow, the generative roles of two kinds of desire, human and divine, stand in sharp contrast (James 1:15, 18, 21):

God's desire,[29]

by means of "his true word," the "word planted deep inside you," "gives us birth . . . and here is the result: we are like the first crop from the harvest of everything he created."

Human desire

"give[s] birth to sin; and when sin grows up, it gives birth to death."

25. See Andrew Chester, "The Theology of James," in *The Theology of the Letters of James, Peter, and Jude,* by Andrew Chester and Ralph P. Martin, New Testament Theology (Cambridge: Cambridge University Press, 1994), 39–41; Walter T. Wilson, "Sin as Sex and Sex with Sin: The Anthropology of James 1:12–15," *Harvard Theological Review* 95 (2002): 147–68 (here, 160–61).

26. See Timothy B. Cargal, *Restoring the Diaspora: Discursive Structure and Purpose in the Epistle of James,* Society of Biblical Literature Dissertation Series 144 (Atlanta: Scholars, 1993), 81–82.

27. Compare *hēdonē* ("pleasure") in James 4:1, 3; *epithymeō* ("I desire") in James 4:2; and *epipotheō* ("I long for") in James 4:5.

28. Cf. Luke L. Cheung, *The Genre, Composition and Hermeneutics of James* (Carlisle, UK: Paternoster, 2003), 206–13.

29. In James 1:18, James uses an aorist passive form of the verb *boulomai* ("I desire" or "I intend"), variously translated as "in fulfillment of his purpose" (New Revised Standard Version), "by his own choice" (New Jerusalem Bible), "he chose" (Common English Bible), New International Version [2011]), "of his own will" (Authorized Version), "he willed" (New American Bible [2011]).

Standing firm leads to the life God promised (James 1:12), but allowing one's cravings to control one's life leads to sin and death (James 1:14–15). As this chart makes plain, over against the power of human craving stands the antidote, the gospel, which when internalized is powerful to save.

Although complex in its presentation, James's argument is clear enough. The challenges of exilic life provide an arena for the unbridled exercise of human desire, the result of which is sin and death. Although one might be tempted to fault exilic life itself, or to lay the blame for seemingly overpowering temptations at God's feet, or the devil's, this is a wrongheaded analysis. The problem is internal, not external, to the human person: human craving, a disposition toward double-mindedness. The solution, though external (that is, God's "true word"), must similarly be internalized. Through a transformation that comes by way of divine wisdom, the divine word must be received and fully embodied so that it imbues who one is and what one does.

Left to their own devices, humans for James are subject to their own cravings; their hearts lean toward disobedience. He does not tell us how this came to be, apart from intimating that this is the way of earthly wisdom. If he provides any hints at all, they come in James 3:6, 15, where we read that "the tongue is set on fire by the flames of hell" and that earthly wisdom is earthy, natural, and "demonic." These references provide no basis for blaming the devil for human unfaithfulness or sin, though, since James apparently thinks that humans can bridle their tongues and sincerely ask for heavenly wisdom. Reflecting on James, Wesley rightly concluded, "We are therefore to look for the cause of every sin, *in*, not *out of*, ourselves."[30] Or as James puts it, "Everyone is tempted by their own cravings; they are lured away and enticed by them" (James 1:14).

Reflections

Before drawing this chapter to a close, let me summarize what we have seen thus far. First, neither Genesis 3 nor scripture as a whole develops much the specific interests that would later coalesce into the traditional doctrine of original sin; that is, scripture does not refer to the Fall, traditionally understood, and nowhere speaks of Adam's sin as a physical inheritance. Second, Jewish literature in the Second Temple period does raise the question of

30. John Wesley, *Explanatory Notes upon the New Testament* (1754; London: Epworth, 1976), 857.

sin's origins, but does not identify sin as an inherent human condition. This literature generally speaks of obedience to God's instructions as the antidote to sin. Third, Paul's more radical view of sin leads him to speak of human servitude to Sin, understood as a power at work in the world, in the face of which humans stand in need of liberation. Simply put, humans need more than God's instructions; they need God's saving intervention in Jesus Christ in order to be freed from sin and enabled to serve God. Fourth, James reflects first-century Jewish thought in his references to "craving" and the evil inclination, but like Paul offers no suggestion that these can be overcome merely through following God's instructions. Instead, James urges that humans need the good news, God's own word, planted deeply inside them. Fifth, when they do reflect on sin's etiology, both Paul and James draw on the opening chapters of Genesis. James reflects on creation in order to assure his audience that the evil inclination, not God, is the author of sin. Paul reflects on Adam both to mark the beginning of the story of humanity's progression into sin and to urge that Adam served as a kind of trailblazer whom people have followed into sinfulness. Both Paul and James thus emphasize sin's corporate dimension and assume sin's heritability—not in the sense of passing sin down through procreation, but in the sense of pattern and influence.

Conclusion

Today's context raises hard questions against the traditional doctrine of original sin. The whole church has never reached a common understanding of "original sin," modern optimism regarding human progress has made it difficult for many to take original sin seriously, theologians have raised ethical objections against the notion that God might hold people responsible for the human sinfulness of past generations, and evolutionary biology has undermined the idea of sin imputed to all of humanity on the basis of the rebellion of our first human parents. Moreover, we have seen that neither Paul nor James affirms that the entire human family is implicated in Adam's sin, and both biblical and extrabiblical Jewish texts typically affirm free will and personal responsibility. What, then, might we say about original sin?

Though it cannot be said that the doctrine of original sin *originated* in the Christian scriptures, we can and should affirm that the doctrine *was developed from scriptural warrants*. This is true at least with regard to our understanding that the human heart leans toward sin, that sin can be intentional or unintentional, that sin is a power that precedes human behavior, and that sin pervades the human family. Moreover, even if Paul and James are not as forthcoming as we might wish on the matter of sin's etiology, both reflect on Genesis 1–3—James to assure his audience that God is not the author of sin and Paul to emphasize the pattern set for all humanity by Adam. On the one hand, this means that there is much of importance to be said for Wesley's defense of the doctrine of original sin on account of the soteriological work it does. On the other hand, the narrative we have thus far traced would be hospitable to an alternative story of "the Fall"—for example, that the lives of our early ancestors were not yet clouded by the haze of spiritual darkness or the muddle of decisions that eventually would envelop the human family as it turned away from God, yet who were subject to temptations and to the desire to turn away from God's voice, and were vulnerable to all kinds of perversions, violence, abuse, and self-centeredness their ancestors bequeathed to them.[31] And such a narrative would be hospitable to a series of elements of the doctrine of original sin—the emergence of sin as a pervasive quality of human experience, sin's personal and structural nature, and sin's character as a disease that pervades the human family relationally (or "environmentally"[32]). Accordingly, on the issue of original sin, scripture provides plenty of room to take seriously the primary questions raised by evolutionary biology. The qualified view of original sin to which scripture bears witness does not require belief in a first human couple, Adam and Eve, or in traditional notions of a historical "fall," or in the traditional view of sin's genetic transmission. The doctrine of original sin—this is not one of those areas where one might urge a forced choice: scripture versus science.

31. Cf. Veli-Matti Kärkkäinen, *Creation and Humanity*, vol. 3 of his *A Constructive Christian Theology for the Pluralistic World* (Grand Rapids: Eerdmans, 2015), 387–411.

32. Anthony C. Thiselton, *Systematic Theology* (Grand Rapids: Eerdmans, 2015), 155–56.

6 The Mystery of Adam

A Poetic Apology for the Traditional Doctrine

AARON RICHES

Nearly every pope since Pius XII has affirmed, with increasing confidence, that there is no conflict between the science of evolution and Christian faith.[1] And yet on the specific question of Adam, the Catholic Church has maintained the traditional view: the Adam of which the Bible speaks is the man who stands at the origin of human history, whose deed against God lies behind the "aboriginal calamity" that set the world "out of joint with the purposes of its Creator."[2] As Pius XII put it in *Humani generis*: the *peccatum originale* "proceeds from a sin truly committed by one Adam (*ab uno Adamo*)."[3] And as the Catechism promulgated in 1992 by John Paul II puts it, citing *Humani generis*: "*Genesis* 3 uses figurative language, but it affirms a primeval event, a deed that took place *at the origin of human history*. Revelation gives us the certainty of faith that the whole of human history is marked by the original fault freely committed by our first parents."[4] The agonizing tension of the church's double commitment—to the origin of human history *ab uno Adamo*, and to the freedom of scientific inquiry that is generally taken by scientists and theologians alike as pointing against

---

1. For key papal statements, see Pope Pius XII, *Humani generis*, 36; Pope John Paul II, Message to the Pontifical Academy of Sciences (22 October 1996), 4; and Pope Francis, Address of His Holiness Pope Francis at the plenary session of the Pontifical Academy of Sciences (27 October 2014).

2. John Henry Newman, *Apologia Pro Vita Sua*, ed. David J. DeLaura, Norton Critical Editions (New York: W. W. Norton, 1968), 248.

3. Pope Pius XII, *Humani generis*, 36.

4. *Catechism of the Catholic Church*, no. 390.

this basic claim—is possible only if the center of unity is Jesus Christ, the "Paradox of paradoxes."[5]

This chapter makes a poetic apology for the traditional account of Adam, understood as the carnal and spiritual origin of human history. I call my appeal "poetic" for three reasons. First, because my argument is aesthetic: it concerns a quarrel about "form" and "narrative," and so in the first place it is not a strictly dogmatic or metaphysical argument (although it implies and serves these). Second, because the chapter drives to an extended commentary (in the last two parts) on Charles Péguy's vindication in poetry of the carnal Adam. Third, and finally, because this chapter pretends to be neither a systematic nor comprehensive account of the debate or its solution, but rather intends to be merely a provocation in favor of a mystery. The apology aims, then, to defend the traditional doctrine of Adam not against or in competition with the scientific evidences, nor yet through a proposed "concordance," but in the light of the tension that shines from the Mystery Incarnate: *in signum cui contradicetur* (cf. Luke 2:34).

I conduct this exploration in six main parts: (1) The exception of theology and two borderline reductions; (2) the historical Jesus and the *figura* of Adam; (3) Christocentrism and Adam; (4) the infrastructure to encounter; (5) the "Dream of Boaz" and the carnal history of Jesus; (6) the generations of Jesus to Adam.

The Exception of Theology and Two Borderline Reductions

The challenge of scientific evidence regarding the traditional view of Adam is ultimately a Christological challenge concerning how the whole inheres and is illumined by the person of Jesus Christ. The concrete mode by which theology may respond to this issue is worked out through what the apostle Paul calls the "analogy of faith" (cf. Rom. 12:6), that is, "the coherence of the truths of faith among themselves and within the whole plan of Revelation."[6] Here the unity of Christ and the unity of the scriptures are mutually constitutive. What is more: since "all things were made through him" (John 1:3), the scriptures are to be understood as intrinsically interrelated with reality in all its factors; and since he "became flesh and dwelt among us"

---

5. Henri de Lubac, *Paradoxes*, in *Oeuvres complètes*, vol. 31 (Paris: Éditions du Cerf, 2010), 8.

6. *Catechism of the Catholic Church*, no. 114.

(John 1:14), then history and biology too must be understood as inter-related in a real way with Christian faith. This last point is crucial: "it is of the very essence of biblical faith to be about real historical events."[7] And this means that the creedal declaration—*Et incarnatus est*—not only declares that God has truly entered into history and into biological flesh, but that the faith of the Bible, this story, is truly history, a dramatic narration of real events enacted by real protagonists.[8] Thus, as Joseph Ratzinger warns, "If we push . . . history aside, Christian faith as such disappears and is recast as some other religion."[9]

Premodern theology presumed without a second thought that the Adam of which the Bible spoke was indeed real: a concrete person, responsible for a concrete deed that became a signature of history, linked to, and illumined by, the far greater signature of history, which is the Paschal Mystery of Christ. And it is as such that Adam is invoked and celebrated in the liturgical life of the church, which does not celebrate abstractions.[10] While the person of Adam is obviously veiled in mystery, in a mythical renarration of the protohistory given in the first chapters of Genesis, nevertheless none of the Fathers, Scholastics, or Reformers thought of him as simply a metaphor or parable.

This truth notwithstanding, on the question of Adam in the face of modern scientific evidence, Christians have tended more and more to fall to one side or the other of two borderline positions, both of which have made their conclusions outside of this unified Christian vision of the whole illumined by the "Paradox of paradoxes."

---

7. Joseph Ratzinger-Benedict XVI, *Jesus of Nazareth*, vol. 1, *From the Baptism in the Jordan to the Transfiguration*, trans. Adrian J. Walker (London: Doubleday, 2007), xv.

8. But of course this is *not* to say that the genre of the Bible is that of a "history book." The Bible is a complex of genres that needs to be read appropriately. On the question of history and genre in the Old Testament, see V. Philips Long, *The Art of Biblical History* (Leicester: Apollos, 1994).

9. Ratzinger-Benedict XVI, *Jesus of Nazareth*, vol. 1, xv.

10. There are many liturgical hymns both of the Greek and Latin traditions that sing of Adam in terms irreducible to a mere idea. The most famous and consequent in the Latin tradition is the *Exultet*, sung at the Easter Vigil, which invokes the "certe necessarium Adæ peccatum, quod Christi morte deletum est! O felix culpa, quæ talem ac tantum meruit habere Redemptorem!" The concreteness of liturgical tradition is substantiated, at a deeper level, when we consider the precedent in both the East and West to celebrate feasts in honor of Adam. Medieval Latin Christians popularly celebrated the feast of Adam and Eve on December 24, and to this day Orthodox Christians celebrate Adam and Eve on the Sunday before Christmas when they celebrate the ancestors of Christ going back to Adam. An "idea" cannot have a feast day.

The first position takes the findings and hypotheses of evolutionary theory and contemporary genetics and begins with Promethean certitude.[11] At least since the publication of Francisco Ayala's "The Myth of Eve: Molecular Biology and Human Origins" in 1995,[12] and overwhelmingly since the 2003 publication of the Human Genome,[13] the best scientific evidence would seem to contradict outright the traditional teaching of the church. The human race, it seems, derived from an ancestral population of no less than 10,000, and certainly not from a single ancestral couple.[14] There cannot, then, have been a single original 'ādām, a lone first human. And neither, moreover, can a sin of origin, a deed, be responsible for the "out of joint" experience of human history.[15] Whatever the link between Christ and Adam in the Gospel of Luke, whatever the doctrine of the apostle Paul, whatever the church's profession for two millennia: Adam can no longer be thought of as a personal agent or protagonist at the origin of human history. The narrative account of Adam in Genesis may be "a poetic and powerful allegory,"[16] but it cannot be that Adam himself is a real figure of personal and historical density.

The second border position is an inverse expression of the first.[17] Rather

11. The most obvious Catholic example of this is evidenced in the work of Pierre Teilhard de Chardin; see for example his *Christianity and Evolution*, trans. R. Hague (London: Collins, 1971). For a more recent Catholic example, see Jack Mahoney, *Christianity in Evolution: An Exploration* (Washington, DC: Georgetown University Press, 2011). For Protestant examples, see Peter Enns, *The Evolution of Adam: What the Bible Does and Doesn't Say about Human Origins* (Grand Rapids: Brazos, 2012); and Denis O. Lamoureux, *Evolutionary Creation: A Christian Approach to Evolution* (Eugene, OR: Wipf & Stock, 2008).

12. Francisco J. Ayala, "The Myth of Eve: Molecular Biology and Human Origins," *Science*, New Series 270, no. 5244 (Dec. 22, 1995): 1930–36.

13. International Human Genome Sequencing Consortium, "Finishing the Euchromatic Sequence of the Human Genome," *Nature* 431 (2004): 931–45.

14. Cf. Dennis R. Venema, "Genesis and the Genome: Genomics Evidence for Human-Ape Common Ancestry and Ancestral Hominid Population Sizes," *Perspectives on Science and Christian Faith* 62, no. 3 (2010): 166–78; Dennis Venema and Darrel Falk, "Does Genetics Point to a Single Primal Couple?," *BioLogos Forum* (April 5, 2010), online at: http://biologos .org/blog/does-genetics-point-to-a-single-primal-couple; and Stephan Schiffels and Richard Durbin, "Inferring Human Population Size and Separation History from Multiple Genome Sequences," *Nature Genetics* 46 (2014): 919–25. But for a decisive account that shows that the scientific evidence need not be interpreted so, see Darrel Falk's chapter in the present volume.

15. Cf. Denis Alexander, *Creation or Evolution: Do We Have to Choose?*, New Edition (Grand Rapids: Monarch Books, 2014), 316–65.

16. Francis S. Collins, *The Language of God: A Scientist Presents Evidence for Belief* (New York: Free Press, 2007), 207.

17. For example, see Victor P. Warkulwiz, *The Doctrines of Genesis 1–11: A Compendium and Defense of Traditional Catholic Theology on Origins* (Mount Jackson, VA: Kolbe Center

than taking the data of scientific evidence as a new Promethean certitude, it takes fragments of the biblical text as self-standing data, "literal truths" convertible with the surest of scientific discoveries, intelligible wholly apart from the *figura* of the whole. Those holding this second position take the Bible as a book of essential data that exist and are "true" on the same plane of reason as the "truths" of modern science. Existing on this univocal plane of meaning, the data of scripture must perfectly "correspond" or "concord" with the "truths" of modern science; if not, it must be that science is in error, since the Bible is inerrant in this reductive sense. The idea in all cases is to re-legitimate Christian faith by insisting that it alone is credible to arbitrate the findings of science according to a predetermined reduction of all reality to the limits of faith conceived as a quasi-modern scientism. Every case of nonconcordance, on this view, is an instance in which science and the Bible are locked in a zero-sum game over the question of a particular truth.

Instead of declaring that "Christ, the new Adam . . . fully reveals man to himself,"[18] as the Second Vatican Council declared, proponents of these two border positions sideline the centrality of the encounter with Christ as the light that must illumine the whole. In the dialectic of dogmatic versus scientific "data," the exception of theology is elided. What is this exemption? That the man Jesus is himself the Logos who sustains the totality of existence in all its perplexity. In him alone reason is held open beyond every reduction, because he is the Forma of the formae of the real, the "Paradox of paradoxes" in whom every factor is saved.

The Historical Jesus and the *Figura* of Adam

The reduction of faith and reason implicit in both our borderline positions on the question of Adam bears an uncanny resemblance to the reductions inherent in that most famous quest of biblical scholarship: the quest for the historical Jesus. While the latter has yielded heretofore undreamt insights into the cultural and historical context of Jesus in which the drama of the Gospels comes to clearer and better light, the quest itself, finally, did not arrive at any definitive or compelling answer to the ultimate question: Who

---

for the Study of Creation, 2007); and William D. Barrick, "A Historical Adam: Young-Earth Creation View," in *Four Views on the Historical Adam*, ed. Matthew Barrett and Ardel B. Caneday (Grand Rapids: Zondervan, 2013), 197–227.

18. *Gaudium et spes*, 22.

is this Jesus? The resemblance between the two quests illumines a few characteristics of our current debate.

In the foreword to the first volume of his *Jesus of Nazareth*, Ratzinger recalls how the quest for the historical Jesus came to a dead-end:

> As historical-critical scholarship advanced, it led to finer and finer distinctions between layers of tradition in the Gospels, beneath which the real object of faith—the figure [*Gestalt*] of Jesus—became increasingly obscured and blurred. At the same time, though, the reconstructions of this Jesus (who could only be discovered by going behind the traditions and sources used by the Evangelists) became more and more incompatible with one another.[19]

It is as if the more focused and exacting the fruits of the historical-critical investigation, the more it definitively set aside both the experience of faith and *sentire cum ecclesia*, the more it became blind to the *forma* who could unite the whole. The incompatibility of the various reconstructions of Jesus formulated by the historical-critical method attest to this: Jesus the anti-Roman revolutionary bore little resemblance to Jesus the sublime moral teacher, neither of which bore much resemblance to Jesus the apocalyptic prophet or Jesus the charismatic healer. And so the quest for the historical Jesus ultimately succeeded in its own self-deconstruction, since what it set out to do was precisely to uncover a putatively lost *figura*. The end of the quest thus left us with a decision: accept that there is a genius inherent in the Christian tradition, that the *figura* of faith handed down for two thousand years indeed illumines the *man* Jesus, or let this Jesus be lost forever to the dustbin of history. Herein lies the analogy with the Adam of Genesis.

While scientific research excels at a breathtaking pace at shedding new and surprising light on human origins, the analogy of faith and the church's tradition has never celebrated Adam as merely a poetical allegory. To rephrase C. S. Lewis's famous "madman or Son of God argument,"[20] let us not come up with any patronizing nonsense about Adam as a powerful metaphor. When truly considered as a whole, the scriptures, the analogy of the faith, and the tradition of the church have not left this open to us. This is not to imply a simple correspondence between the historical Jesus and the historical Adam, as if the two "quests" were equivalent. Clearly

19. Ratzinger-Benedict XVI, *Jesus of Nazareth*, vol. 1, xii.
20. C. S. Lewis, *Mere Christianity* (London: Collins, 1952), 54–56.

in the first case we are dealing with a fact of recorded history: there was a man named Jesus of Nazareth, just as there was a man named Alexander the Great. The Christian claim concerns the meaning of this fact; the historical life of this Jesus belongs so entirely to the divine Logos as to be personally constituted by him, so as to be "one" with him. In the history of this first-century Palestinian Jew, thus, the Logos has submitted himself to history, and so to historical scrutiny. In the case of the latter, the historical Adam, we are dealing with something radically different. Adam is not God, he is the first being of human history. The Christian claim here is that, with Adam, we have the birth of human history itself, which means, paradoxically, that "Adam" is anterior to "history," in the sense that he is a figure of the proto-logical past and as such cannot be investigated by historical methods. Nevertheless, in both cases, the case of Jesus and of Adam, the wager of faith involves more than what history or science can establish: it involves a radical trust in the genius of the tradition, what has been "handed on" (cf. 1 Cor. 11:23; 15:3–8). In the case of Jesus it involves the wager that he is the Son of the Father, and Savior of the world. In the case of Adam it involves the wager that, as Hans Urs von Balthasar puts it, the "Adam" of Genesis was "not *any* individual, but the one who founded the family of mankind," and that by his "decision against God," he has "plunged this whole family, not into personal sin, but into a lack of grace."[21] Adam is either this original protagonist of history and of the calamity of sin—however veiled in mystery—or we must allow him to be swept into the dustbin of, in this case, protohistory.

Christocentrism and Adam

How then are we to proceed? How do we establish a meaningful *figura* of Adam in the face of the scientific evidence? Paradoxically, we ought not to begin with him, but with the *figura* of faith, Jesus Christ. This is to insist on thinking theologically about Adam, that is, contemplating him within the analogy of faith, the "coherence of the truths of faith" that lie at the incarnate core of the Christian claim. And this is where the apostle Paul begins in Romans 5: Adam *est forma futuri* (5:14). Adam is the *typos*, the "type," the *figura* of "the one who was to come" (5:14).

---

21. Hans Urs von Balthasar, *Theo-Drama*, vol. 4, *The Action*, trans. G. Harrison (San Francisco: Ignatius, 1994), 183.

Paul's formulation suggests an analogy: Adam is like a wax impression in relation to the iron stamp who is Jesus Christ. Adam is the type; Christ is the archetype. And this means that the wider context of Romans 5:12–21, which furnishes the biblical basis of the classical doctrine of *peccatum originale*, concerns first of all—in a metaphysically prior sense—the question of the person of Jesus and his redemptive deed, and only then, in a metaphysically secondary sense, the person of Adam and his decision against God. This is the conclusion of Pope Benedict XVI, who in his catechesis on Romans 5:12–21 says that "the center of the scene . . . is not so much Adam, with the consequences of his sin for humanity . . . as it is Jesus Christ and the grace which was poured out on humanity in abundance through him."[22] Only in the grace that pours forth from the Crucified One does the Fall of the first man become concrete to us. And so: only in the greater *figura* of Christ does the *figura* of Adam come to light; only in the redemptive deed of the cross does the meaning of the deed at the beginning of human history become clear; only in the grace of the crucified does the fall of the first man become real to us as a concrete fact we can (and must) recognize. It is by contemplating the historically particular person of Jesus Christ and the event of his Paschal Mystery that the prior density of Adam and his deed is illumined.

Yet as soon as we situate this priority in Christ we must emphasize that this nevertheless requires, on Christological grounds, a concrete Adam, irreducible to a mere "idea." Adam is a type as real and concrete as his archetype, who therefore cannot be reduced to a mere concept or metaphor. As N. T. Wright claims, "Paul clearly believed that there had been a single first pair, whose male, Adam, had been given a commandment and had broken it."[23]

The Pauline doctrine offered in Romans 5 fully presumes an incarnate correspondence between the historical fact of Christ and the concrete mystery of Adam. This correspondence is key to how the saving work of Christ functions, how humanity is transferred from being dead "in Adam" to being saved "in Christ." Adam is a "particular person . . . who stands at the origin of humankind and with whom the history of sin begins."[24] It is precisely as such—as the person at the origin of human history and the protagonist of the sin of origin—and

---

22. Pope Benedict XVI, *Saint Paul: General Audiences (July 2, 2008–February 4, 2009)* (San Francisco: Ignatius, 2009), 90.

23. N. T. Wright, "Romans," in *The New Interpreter's Bible*, vol. 10 (Nashville: Abingdon, 2002), 526. Further on Paul's "historical" Adam (with bibliography in the notes), see C. John Collins's nonconformist apologetic, *Did Adam and Eve Really Exist?* (Wheaton, IL: Crossway, 2011), 78–90.

24. Ratzinger, *"In the Beginning . . ."* (Grand Rapids: Eerdmans, 1995), 71.

not otherwise, that Adam is "a type of the one who was to come" (Rom. 5:14). The carnal and historical density of Adam is known as a concrete reality because he is a type of Christ, the most carnal and historical of all human beings.

The Christocentrism Pope Benedict highlights in Romans 5 is articulated more radically by Conor Cunningham in the last chapter of his book *Darwin's Pious Idea*, in which he sets out to un-ask a series of metaphysically reductive questions around the "historical Adam." Paul's strategy, as Cunningham deploys it, destabilizes the modern conception of time/history as strictly linear and being/existence as purely punctiliar. History is not a linear series, neither are beings an aggregate of discretely existing "substances." Why is this? Because in Trinitarian terms, the persons of the Trinity are "subsistent relations," henceforth "*relatio* . . . stands beside . . . substance as an equally primordial form of being."[25]

On the one hand, then, in light of Trinitarian theology, the relationship between different beings must now be understood not merely as "accidental," but as a constitutive factor of their substantial reality. Every child, for example, has a mother and father, who give relation a substance in the form of the child that emerges from them. If this is true of a relation between a human child and parent, it is more so of the creature's relation to God, who is not merely the atemporal origin of created being but its sufficient present reality. The creature only exists as a being-in-relation to God. In addition, the recovery of relation as a primordial aspect of being helps us to understand better the truth of historical being: no event of historical being can be understood in isolation, but to the contrary, is always implied and so must be understood within the greater web of interrelation that is the whole of history. And if the foregoing is the case generally, it is overwhelmingly so with Adam, who is tasked by God to be the father of the human race (the carnal source of all human being), and so to represent the totality of the human being in relation to God at the origin of human history.

Adam is the universal man, the *principium* of human history and being. As such he is the figure, the type, of the true *principium*, Jesus Christ, who is the archetype of human being and whose life is the still point around which all history turns. Thus Cunningham enjoins us to follow Paul: "we should think of the 'type' as having reality only in that of which it is a type."[26] The

---

25. Joseph Cardinal Ratzinger, *Introduction to Christianity*, trans. J. R. Foster (San Francisco: Ignatius, 2004), 183.

26. Conor Cunningham, *Darwin's Pious Idea: Why the Ultra-Darwinists and Creationists Both Get It Wrong* (Grand Rapids: Eerdmans, 2010), 378.

deepest reality of Adam, then, lies wholly in Christ, just as the deepest truth of his original sin points already to the glory of the Cross. And so, "We only understand Adam because of the one true Adam . . . the *only* Adam. . . . [Which means it] is folly to interpret the [historical event of the] Fall or the existence of Adam in either positivistic terms or strictly historical [as in "historicist"] terms, in the sense that there is no Fall before Christ."[27] But does this radical Christocentrism entail a Christological reduction? Does figuring Adam in this way collapse his whole evental and personal reality into Christ? In a recent defense of the traditional doctrine of Adam, Hans Madueme reads Cunningham's argument as ending up precisely in this kind of Christo-monism: "Adam is swallowed up by Christology and soteriology."[28]

Madueme is right to warn against Christo-monism. A Christological reduction, tempting as it might be, has to be rejected for Christological reasons, since Christ illumines the ultimate value of being as lying in its co-extensiveness with love. Hence Christology entails, from its origin, an incarnate relationality between concrete agents. "Relation" is therefore as primordial to his incarnation as "substance," so that the substantial relation of the incarnate Lord presupposes and entails the freedom of the genuine "other."[29] This is what is meant by the claim that Jesus "fully reveals man to man himself," not by himself, that is, not as a being apart, but rather Jesus "reveals man to man himself" in the greater "revelation of the mystery of the Father and His love." The revelation of the Father and his love—which occurs in the concrete carnality of human flesh received by the Son—brings to light the ultimate valuation of human being in the plan of divine love. God creates the human being to be a communion partner, a real co-agent in the work of love that is the deepest truth of God's own being. This is the Christo-logic of the world that disallows every Christological reduction of the world (and every human being and human being as such) to Christo-monism. Christ reveals the human to himself only when he reveals the Father and his love. Everything points, thus, to the dialogue of love: the communion of real personal agents who are genuine agents of love's communication. Madueme's insistence that Adam not be reduced to Christ is Christologically expedient.

27. Cunningham, *Darwin's Pious Idea*, 378.
28. Hans Madueme, "'The Most Vulnerable Part of the Whole Christian Account': Original Sin and Modern Science," in *Adam, the Fall, and Original Sin: Theological, Biblical, and Scientific Perspectives,* ed. Hans Madueme and Michael Reeves (Grand Rapids: Baker Academic, 2014), 225–50, here at 231.
29. See Adrian J. Walker, "The Original Best: The 'Coextensiveness' of Being and Love in Light of *GS*, 22," *Communio* 39 (2012): 49–65.

But his charge against Cunningham misses the mark to the extent that it undervalues the role of paradox Cunningham presumes in the relation of "type" to "archetype." Let us explore this further.

In a critically important discussion of the relationship between Old Testament sacrifice and the sacrifice of Jesus Christ, the author of the Epistle to the Hebrews declares: "the law has but a shadow of the good things to come instead of the true form of these realities" (Heb. 10:1). Commenting on this passage, Henri de Lubac highlights just "how daring" the expression is: it entails a "complete reversal" of "common sense," since we normally think of the archetype or model as preceding the type, just as a cause precedes its effect.[30] But here the "rough sketch is the preparation for the archetype" and so the imitation impossibly comes before the actual model. De Lubac goes on:

> This is the truth that is to come and will arise one day on earth: *futura Veritas, secutura Veritas—Veritas de terra orta.* Unheard of paradox! . . . [This] is the disconcerting reality of the Christian fact; it is both the substance and model, the truth that is foreshadowed and reflected in . . . what went before. . . . The whole Christian fact is summed up in Christ . . . who [nevertheless] had to be prepared for in history, just as a masterpiece is preceded by a series of rough sketches.[31]

All of this is entailed in the metaphysical fact that in God there is no "before" and no "after." The incarnation of God both reconfigures the whole future course of history, while it recapitulates the entirety of the past. Only because this is the case can the church declare: "O certe necessarium Adae peccatum . . . O felix culpa, quae talem ac tantum meruit habere Redemptorem!" John Paul II says as much when he declares that Jesus Christ is "the centre of the universe *and of history.*"[32] Metaphysically speaking, the Christ event is the first event, the event from which, to which, and for which all of human history unfolds. But to claim that Christ is the archetype, the *Forma* in which the *forma* of Adam comes to light, is in no way to reduce history to the incarnation, or Adam to Jesus. Every reduction of Adam—whether to Christology, to a metaphor, or an abstraction—is dubious on Christological

---

30. Henri de Lubac, *Catholicism: Christ and the Common Destiny of Man*, trans. Lancelot C. Sheppard and Sister Elizabeth Englund, OCD (San Francisco: Ignatius, 1988), 172.

31. De Lubac, *Catholicism*, 173–74.

32. John Paul II, *Redemptoris hominis*, 1. My italics.

grounds. The Christian claim is at its core incarnate, personal, and historically concrete, an in-breaking of God in the middle of time. It presumes that
God in Christ received his incarnate and historical being from a preexisting
and carnal history. The history of Jesus Christ has a carnal root. That carnal
root is the first man, who is the *figura* of the *Figura* to come.

The Infrastructure to Encounter

N. T. Wright's claim is not merely that Paul clearly presumed that Adam was
the first man, who had received a divine commandment he transgressed;
he believed "that there had been a single first pair."[33] This would seem to be
the view of Jesus himself, as he responds to the Pharisees who came to test
him by asking if it is ever lawful to divorce one's wife (Matt. 19:3–9; Mark
10:2–9). Jesus answers their question thus:

> "Have you not read that he who created them from the beginning made
> them male and female [cf. Gen. 1:27], and said, 'Therefore a man shall
> leave his father and his mother and hold fast to his wife, and the two
> shall become one flesh' [cf. Gen. 2:24]? So they are no longer two but
> one flesh. What therefore God has joined together, let not man sepa-
> rate." (Matt. 19:4–6)

Jesus's response ties together the two accounts of man and woman (Gen.
1:27; 2:24) in a way that clearly implies that he understood the two accounts
as referring to one and the same human couple.[34] And more, this unity of the
creation of the first man and woman is the basis of his more radical answer to
the Pharisees' wish to justify divorce based on the command of Moses. Jesus
responds: "Because of your hardness of heart Moses allowed you to divorce
your wives, but from the beginning it was not so" (Matt. 19:8). As Pope John
Paul II expounded with beauty and rigor in his catechesis on human love,
the ontological density of Jesus's argument, "from the beginning," cannot
be underplayed.[35] The metaphysical constitution of the first pair, in their

33. Wright, "Romans," 526.

34. John Collins places great importance on this discourse as foundational of the New
Testament view of Adam and Eve; see his comments in *Did Adam and Eve Really Exist?*,
76–78.

35. Pope John Paul II, *Man and Woman He Created Them: A Theology of the Body*, trans.
Michael Waldstein (Boston: Pauline Books & Media, 2006).

complementarity, leads us to the deeper infrastructural meaning of human existence as ordered to love, as ordered to the personal awakening of the human being to the depth of the mystery of his humanity through the joyful surprise of encountering the other.

Because it was so "from the beginning," because it was so from Adam and Eve, every encounter of an "I" falling in love with a "you" is a revelation of the mystery of being human as ordered to the ultimate encounter of the "I" with the incarnate Christ, who "fully reveals man to himself and brings to light his most high calling."[36] Christ is the unity of the beginning with the end; he is the Alpha and the Omega. This implies that the *forma* of the archetype, Jesus, contains within himself, as the *forma* of his own being, the infrastructure of personal encounter, such that he condescends to require a real personal history that truly precedes him and cannot be reduced to him, in order that his coming may be an encounter of love—and this is the way in which he enfolds all of reality in himself, not as a reduction, but rather in the form of the recognition of a genuine other.

This theological point of departure finds an antecedent correlate in the carnal story of Adam, in what John Paul termed "the bone marrow of the anthropological reality."[37] This bone marrow of anthropological reality is the first joyful cry of human history, that of Adam upon the creation of Eve: "This at last is bone of my bone and flesh of my flesh" (Gen. 2:23). According to John Paul the first man "speaks these words as if it were only at the sight of the woman that he could identify and call by name . . . *that in which humanity is manifested*."[38] The epiphany of this encounter is crucial. Only by awakening to the presence of Eve does Adam become conscious of the meaning of his own humanity.[39] She reveals him to himself. It is as if the infrastructure of the human being was created precisely for this encounter. This original encounter, surely as significant as the Fall in the biblical narrative, unwaveringly points us to the encounter with a God as the Beloved who will become at last bone of our bone and flesh of our flesh and reveal us to ourselves. It is from this original cry of joy that the whole genealogy of human history and biology radiates, now as a call to pursue something beyond appearances, something both within and beyond the data of science. This is, in fact, the

36. *Gaudium et spes*, 22.

37. John Paul II, *Man and Woman He Created Them*, 9.4, p. 164.

38. John Paul II, *Man and Woman He Created Them*, 9.4, p. 164.

39. In *Denial: Self-Deception, False Beliefs* (New York: Hachette, 2013), Ajit Varki and Danny Brower offer a correlate—if wholly negative—account of the uniqueness of the human brain in comparison to all other animals.

elemental human experience—that of the wonder of the human being before reality, one that spurs on new advances in scientific investigation and provokes the disposition of adoration before the Mystery of being. It is summed up in some lines of the Italian poet Leopardi, who writes:

E quando miro in cielo arder le stelle;
Dico fra me pensando:
A che tante facelle?
Che fa l'aria infinita, e quel profondo
Infinito Seren? che vuol dir questa
Solitudine immensa? ed io che sono?[40]

This is the vital élan of the human spirit that animates every human adventure. This human pondering, this striving toward the luminous encounter with Christ, who is himself, as John Paul II called him, the "answer to the question that is every human life."[41]

The Dream of Boaz and the Carnal History of Jesus

In pursuit of this reality beyond immediate appearances, I want to explore a proposal of the French Catholic poet Charles Péguy.

In the introduction to a recent Spanish edition of Péguy's work, Javier Martínez and Sebastián Montiel characterize the French poet as a twentieth-century "prophet of the Incarnation."[42] According to them, Péguy's poetry, as the substance of his prophecy, bears "witness to [the unity] of being

---

40. ". . . as I / Behold in heaven the fiery stars, I ask myself / Why so many blazing torches? / What's the point of the endless air / Or the infinite deep reaches of sky? / What does this huge solitude mean? Or what am I?" Giacomo Leopardi, "Night Song of a Nomadic Shepherd in Asia," in *Leopardi: Selected Poems*, trans. and ed. Eamon Grennan (Princeton: Princeton University Press, 1997), 57–66, at 61. Translation slightly modified. Cf. Luigi Giussani, *The Religious Sense*, trans. John Zucchi (Montreal: McGill-Queen's University Press, 1997), 45–46.

41. John Paul II, Homily at Camden Yards, Baltimore, October 8, 1995, as quoted in George Weigel, "Diognetus Revisited, or, What the Church Asks of the World," in *Against the Grain: Christianity and Democracy, War and Peace* (New York: Crossroad, 2008), 64–84, at 76.

42. Javier Martínez and Sebastián Montiel, "Introducción," Charles Péguy, *El Frente está en todas partes (Selección de textos)*, trans. and ed. Javier Martínez and Sebastián Montiel (Granada: Nuevo Inicio, 2014), 55.

and event," that is, the irreducible coinherence of history and metaphysics, fact and existence, that marks the concrete experience of human reality and is the basis of Christian orthodoxy. This basis of theology, however, is not something Péguy learned from the schoolmen of his day, but rather from the work of his poetical hero Victor Hugo, a poet known for his anti-Catholic views (in his last will, Hugo famously refused the "orations of all churches," while begging the "prayer of all souls").

In a 1910 essay dedicated to Hugo and what he learned from him, Péguy highlights Hugo's poem "Booz endormi" (Boaz Asleep), a poetic interpretation of a scene from the book of Ruth.[43] In "Booz endormi" Péguy discerns a critical "pagan witness" to the incarnation as the central fact of human reality. The key verse of Hugo's poem reads as follows:

> Et ce songe était tel, que Booz vit un chêne
> Qui, sorti de son ventre, allait jusqu'au ciel bleu;
> Une race y montait comme une longue chaîne;
> Un roi chantait en bas, en haut mourait un dieu.[44]

This verse clarifies, for Péguy, a vision of human experience that awakens us to the mystery of the incarnation that grows out of human flesh. The singing king, the vision of joy and power, stands at the root, supporting the dying god who is the summit of the flourishing of human life. On one level, the god who dies is the Logos, the ultimate ground of all that has being, yet the god receives his dying flesh as the ultimate flourishing of what he has made, and this is the particularly pagan aspect, this attention to the god *receiving* his flesh, the ground of his carnal being, almost as if humanity produces the incarnation.

For Péguy, what is central to this "pagan witness," to this true light shed on a sometimes forgetful Christianity, is that the incarnation surges on one level from the urgent cry of the human flesh. Orthodox Christian faith takes its point of departure from the other side: in both the Nicene Creed and in

43. Charles Péguy, "Victor-Marie, comte Hugo," in *Œuvres en prose complètes*, vol. 3 (Paris: La Pléiade, 1992), 161–345. I have also used the Spanish translation of selections of this essay by Martínez and Montiel, titled "La encarnación" in Péguy, *El Frente está en todas partes*, 155–61.

44. "And from his loins a great oak, flourishing, / Stirred Boaz in his dream, and, gazing down, / He saw a race ascending it; / A king sang at the roots; a god died in its crown." Victor Hugo, "So Boaz Slept," trans. R. S. Gwynn, in *Poets Translate Poets: A Hudson Review Anthology*, ed. Paula Deitz (Syracuse, NY: Syracuse University Press, 2013), 82–85. These four lines are quoted by Péguy in "Victor-Marie, comte Hugo" at 236.

the Gospel of John, the incarnation is overwhelmingly established, not by the upward longing of the human heart, but by the descent of the Son of God, who "came down from above" (*descensus de caelis*). In Hugo's poem, however, Péguy reads an apology for the original, elemental experience of the human being, that which is constituted in all of its aspects by a carnal urgency to give fleshy birth to the divine mystery.

Thus, if the proper starting point of Christian theology and faith is the encounter with the divine Logos who came down from above (cf. John 6:38), then what the pagan witness helps faith to remember is that there are two movements that meet in the one Son: he is eternally "of the Father" (*ex Patris*), but nevertheless he is also carnally "of Mary" (*ex Maria*). Therefore Jesus has a genealogy: a flesh and history that ensures that he is truly "made man" (*homo factus est*), and so proceeds from the human.

Péguy takes hold of the tension that this double movement of the incarnation implies in order to show that, while the incarnation can be contemplated, as it normatively is by theology, in the *figura* of Jesus who came down from above, there is a "pagan" correlate contemplation. Instead of contemplating Christ from the perspective of the eternal (*ab aeternitate*) as "a history that has happened *to the eternal*,"[45] the incarnation can be contemplated as a mysterious correspondence of a happening within flesh itself, as "the carnal plenitude of a carnal series."[46] And it results in a history of Jesus understood "as a history *which has happened to the soil* that HAS GIVEN BIRTH TO GOD."[47] From this angle the temporal is understood as an event of flowering, and the incarnation a temporal fruit, indeed "a fruit of the earth . . . as a history (a consummating, supreme, limit) happening to flesh and to the earth."[48] Péguy writes: from this angle we consider and contemplate God legitimately "from the side of his creature, as coming from the side of his creature [from *inside* his creature] . . . God coming into his creature, the creature welcoming (his) God, through a series of creatures: the lineage of David, leading to God as a carnal fruit."[49] The flesh of Israel, on this view, produces the incarnation in the sense that it truly desires—in a carnal way—for God to be, at last, bone of her bone and flesh of her flesh. Rightly understood, this upward human longing cannot in any way of course determine the descent

45. Péguy, "Victor-Marie, comte Hugo," 235.
46. Péguy, "Victor-Marie, comte Hugo," 234.
47. Péguy, "Victor-Marie, comte Hugo," 235. All italics and capitals in Péguy's text are his.
48. Péguy, "Victor-Marie, comte Hugo," 235.
49. Péguy, "Victor-Marie, comte Hugo," 236.

of God into flesh or diminish the surprise of the total divine initiative, much less demand it. Theology will rightly insist that the creature is the creature and God is God. Nevertheless, what this other mode of contemplating the incarnation helps us to remember is this: the utterly concrete and carnal specificity required for any adequate theology of the incarnation.

The Generations of Jesus to Adam

The ultimate verification of the "pagan witness" of the incarnation, according to Péguy, comes from the tensive complementarity of the Matthean and Lucan genealogies. The former is, according to him, normatively Christian—it is concerned with election and salvation, and with the origin of faith—while the genealogy of Luke, the Evangelist of the Gentiles, is "pagan" in the sense that it is concerned with rootedness in the soil, with original flesh, and with the carnality of Jesus as a fact traceable to the ultimate source, to the human original.

In truth, according to Péguy, the Matthean genealogy is not a genealogy but an account of the generations of Jesus from Abraham, "who was the second Adam."[50] What does it mean to call Abraham the second Adam? For Péguy, it means that he is not only carnal, created, tempted, and cast out—as the first Adam was—but he is also carnally born, carnally elected, and carnally chosen to be the father of the elect people: an origin that is "carnal and spiritual at once."[51]

To the carnal-spiritual origin of faith in election, there is for Péguy a second key, indicated by the narration of generations that pass through concrete "crimes of the flesh."[52] These crimes are signaled by the invocation of at least three of the four women mentioned in the genealogy: Tamar, Rahab, and Bathsheba ("the wife of Uriah"). The names conjure up the most carnal and horrible sins: rape, prostitution, adultery, murder, incest. The legacy of the first Adam in this way echoes through the genealogy that begins with the election of Abraham, the father of faith. Péguy writes: "It has to be recognised, the carnal lineage of Jesus is frightful. Few men, few *other* men, could possibly have had so many criminal ancestors, and so very criminal,"

---

50. Péguy, "Victor-Marie, comte Hugo," 237.
51. Péguy, "Victor-Marie, comte Hugo," 237.
52. Péguy, "Victor-Marie, comte Hugo," 238.

as did Jesus.[53] Burdened by the shame and calamity of sin, the carnal and criminal lineage of Jesus is, according to Péguy, internal to the meaning of the divine descent and "confers on the mystery of the Incarnation its whole value."[54] And Matthew hides nothing.

While Matthew's "generations of Jesus" roots the incarnation in faith and in election, the Lucan genealogy offers something different: it moves in the opposite direction, searching from Jesus backwards to the carnal origin.[55] While Matthew begins his Gospel with the generations of Jesus, Luke waits until the beginning of Jesus's ministry before he searches out the genealogy of Jesus, "the temporal race . . . [that goes] all the way back to the first Adam, the carnal Adam."[56] Whereas Matthew descends through time, Luke rises through it. It is as if in both cases, Péguy reminds us, we are dealing with the same time, only one coming down and the other going up. Whereas Matthew roots his genealogy in Abraham our father in faith, Luke roots his in Jesus himself: "Jesus, when he began his ministry, was about thirty years of age, being the son (as was supposed) of Joseph" (Luke 3:23). From here Luke rises: step by step, son to father, son to father, to "the son of Enos, the son of Seth, the son of Adam, the son of God" (Luke 3:38).

At its heart, according to Péguy, the point of the upward scaling from son to father—*qui fuit, qui fuit*—from carnal generation to carnal generation, is to arrive at the first Adam and the carnality of the first Adam, the son of God, received from God himself. The mystery of the carnal filiation of Adam is nothing less than that of the incarnate Son precisely in the fact that it equates Adam's carnal generation from God in just another *qui fuit*. And, so for Péguy, the mystery of the carnal filiation of the first Adam signals that "there is something carnal in *Pater noster*, in *Our* FATHER."[57] God the Father is a true and carnal father to Adam, a flesh-giving father. God is a father to Adam in the same way any father is the father of any son. The only thing that distinguishes every other human father from God is that behind God there is no other. The doctrine of monogenesis is true: all humanity springs from a single parent, the Father of Adam.

53. Péguy, "Victor-Marie, comte Hugo," 239.
54. Péguy, "Victor-Marie, comte Hugo," 239.
55. Péguy, "Victor-Marie, comte Hugo," 240.
56. Péguy, "Victor-Marie, comte Hugo," 240.
57. Péguy, "Victor-Marie, comte Hugo," 242.

Conclusion

I began this chapter with the suggestion that whatever the unbearable tension between the traditional doctrine of Adam and the current consensus of science, theology ought not to set off impatiently after a "synthesis," but rather should risk abiding the Paradox of paradoxes, Jesus Christ. The inscrutable mystery of the origin of human history is betrayed by a theological imagination overly determined by contemporary science, whether this takes the form of a tortured conceptual synthesis, the reduction of Adam to a metaphor, or in the creationist rejection of the scientific evidence. The origin of human history is doubly veiled by the overdeterminations of the divine mystery and by the "aboriginal calamity." And yet God himself sanctions the search for Adam.

After the sin of origin, as the luminously dark text of Genesis narrates it, when "the man and his wife hid themselves from the presence of the LORD God among the trees of the garden" (Gen. 3:9), God went out in search of his creature. "[W]alking in the garden in the cool of the day . . . the Lord God called to the man and said to him, 'Where are you?'" (Gen. 3:8-9). The restless cry of God sets in motion the divine abandonment of descending Love that finds its ultimate echo in the human cry of the crucified Son: "My God, my God, why have you forsaken me?" (Mark 15:34). The drama of human history unfolds between the cry of God looking for Adam and that of Christ calling out for God. But the divine descent does not stop here: it plunges God himself to the bottom of hell in search of his fallen creature. Only there, at the furthest limit, does the God-forsaken Son find the first human being, Adam, in the far country of his own God-forsakenness. As the ancient Greek icon of the harrowing of Hades depicts Jesus pulling Adam and Eve out of the void of death, so Latin Christians in the Middle Ages celebrated, on December 24, the feast of the salvation of Adam and Eve. Christ is not only the revelation of what it means to be human; he is the unbridgeable crossing of the chasm that separates sinful humanity from the face of its living origin, the first Adam. Only Christ saves the human being—at his deepest carnal root—from the pit of oblivion.

PART III

Beyond "Origins": Cultural Implications

7 Being All We Should Have Been and More

The Fall and the Quest for Perfection

BRENT WATERS

The Fall, as recorded in Genesis 3, is a story that pervades scripture. God places Adam and Eve in the Garden of Eden with the charge to care for it. In fulfilling this responsibility, they are permitted to eat from any tree in the garden, particularly the tree of life. But God forbids them to eat the fruit of the tree of the knowledge of good and evil, death being the penalty if they disobey this prohibitive command.

The serpent tempts Eve, suggesting she will not die if she eats the forbidden fruit. Rather, she will be like God who is simply trying to protect a kind of divine monopoly on knowledge and wisdom. Eve and Adam succumb and eat the prohibited fruit. The consequences are disastrous. They are cast out from Eden, denying them access to the tree of life that sustains them. Since they now know good and evil they must not be allowed to live forever, and they eventually die. Their relationship with God is broken; no longer do they walk with God in the garden. Their relationship with each other is ruined; equality is displaced by Eve being subjected to the rule of Adam. Even nature turns against them; gardening is no longer an effortless pleasure but a curse of endless drudgery. Moreover, subsequent generations suffer the results of Adam and Eve's disobedience.

Whether one interprets this story literally, symbolically, metaphorically, mythologically, or whatever other term might be added to the list, the basic storyline is that the human condition is not as it should be and is in need of correction. In eating the fruit of the tree of the knowledge of good and evil, Adam and Eve consign themselves and their descendants to a discordant and sinful life of disordered desire resulting from their broken fellowship with

God, creation, and their fellow creatures. This theme is repeated time and again throughout the Bible. If humans cannot reenter the Garden of Eden, they will construct their own, suitable substitute. We are told, for instance, in Genesis 11 that humans gather at Shinar to build a city with a tower to reach the heavens in order to make a name for themselves and no longer be wanderers in the inhospitable world in which God has exiled them. They will seek to be like gods as their ancestors, Adam and Eve, rightfully strove to become. But God will not allow any striving for a godlike status, and confuses their language, effectively scattering them once again across the earth. Much of the Bible can be characterized, in part, as an account of humans pursuing one failed attempt after another in trying to overcome the Fall on their own terms, and the strife and suffering resulting from their futile efforts.

To say that the Fall, particularly when conjoined with the pagan notion of a lost Golden Age, has profoundly influenced the Western religious, moral, and intellectual imagination is an understatement.[1] Christianity, for instance, is inconceivable without some understanding of sin from which humans must be saved. Morality or ethics would not be needed if human desire were properly ordered rather than disordered. Individuals and societies would not suffer the ill effects of envy, avarice, and vainglory if their fellowship had remained unbroken. The need for political coercion to maintain a modicum of civility would have no role to play if the original harmony of Eden could be recovered. If only Eve and Adam had not eaten the forbidden fruit and presumed a godlike, though counterfeit, knowledge of good and evil, the human condition would not be such a mess.

But it is a mess, and some notion of the Fall, in both its religious and secular guises, presents a challenge to be overcome, namely, how might human nature be restored, improved, or even perfected? A variety of religious and secular responses to this challenge have been proffered, three of which may be noted for illustrative purposes.

First, the perfection of human nature is an eschatological hope. This side of the *parousia*, humans will remain fallen creatures. Since they have only a vague or distorted understanding of good and evil, any attempts at improving the human condition will prove fruitless or worse. Consequently, humans should wait patiently until God completes the redemption of creation. Such hope, however, may inadvertently promote fatalism, inspiring

---

1. See, e.g., Peter Harrison, *The Fall of Man and the Foundations of Science* (Cambridge and New York: Cambridge University Press, 2007); see also Terry Otten, *Visions of the Fall in Modern Literature* (Pittsburgh: University of Pittsburgh Press, 1982).

moral indifference in respect to material wants and needs, or worse, gnostic disdain of the world and its less enlightened inhabitants.

Second, the perfection of human nature can be achieved through will-power. Although humans are fallen creatures, they nonetheless retain a capacity to clarify their understanding of good and evil. As the Pelagian gloss on the rich young ruler (Matt. 19:21) makes clear, there are certain things one can do to become perfect.[2] Humans have both the potential power and responsibility to correct the ill consequences of past deeds if only they will themselves to do so. In short, the human condition can be improved, even perfected, if the will is aligned with the good. Such confidence in a clear perception of what the good is and the strength of will to do it, however, can result in a prideful and bigoted zealotry. When one is intoxicated with a clear understanding of the good, it is difficult, perhaps even irresponsible, to be patient with those lacking such clarity or possessing too weak a will to become perfect.

Third, the perfection of human nature can be engineered. Similar to what Pelagians thought, humans retain the capacity of reason to clarify the knowledge of good and evil. Yet unlike with the Pelagians, there is little confidence in willpower, but weak wills can be manipulated into doing the good by creating suitable contexts in which their wills are asserted. Through appropriate moral, social, political, and biological engineering, the human condition can progressively be improved, perhaps even perfected over time. Such confidence in technological mastery, however, can create a myopic intolerance particularly when accompanied by religious, moral, or ideological convictions. Those standing in the way of progress must either be marginalized or eliminated in order to correct the human condition.

As indicated above, these are merely three illustrative themes in a broader history of the quest for perfection. John Passmore depicts this sordid history in both its religious and secular manifestations.[3] How human perfection is defined and how it should be achieved has varied over time and across social locations. Personal virtue, collective equality and harmony, and evolutionary progress, for instance, have been championed as ideals to be pursued. The ways for achieving them have been propounded, respectively, as philosophic contemplation, social ordering, and biological and genetic engineering. The results of these efforts, however, have been far from per-

2. See "On Riches," in B. R. Rees, *Pelagius: Life and Letters* (Woodbridge, UK: Boydell, 1998).

3. John Passmore, *The Perfectibility of Man* (New York: Charles Scribner's Sons, 1970).

fect. The contemplative life of philosophers requires a vast supportive infrastructure of exploited laborers or slaves to provide such vulgar necessities
as clothing, shelter, sustenance, and the like; virtuous perfection requires
that the many be consigned to serving the needs of the elite few. Achieving
egalitarian perfection requires a radical reordering of social relationships
in which the powerful are brought down and the weak uplifted; gulags and
purges are unfortunate and temporary necessities for achieving the perfect
society. Natural selection is too slow and cumbersome, so humans should
direct their evolution toward the goal of biological perfection; eugenics is a
hygienic step in culling, and thereby improving, the gene pool. As Passmore
demonstrates, the quest for perfection, however defined, invariably ends
with intolerance and inhumane treatment for those unable or unwilling to
attain the proffered perfection.

It has been more than forty years since *The Perfectibility of Man* was first
published, so it does not include the latest quest for perfection. This quest is
more immodest than its predecessors, for the goal is not merely to perfect
humans but to redirect evolution in order to make them better than human,
to create the posthuman. The term "posthuman" refers to a wide-ranging
discourse on the prospect of directing future evolution toward the goal of
creating new and superior beings.[4] The most prominent proponents of this
ambitious project are the self-identified transhumanists, represented by such
authors as Nick Bostrom, Max More, Aubrey de Grey, James Hughes, Hans
Moravec, and Ray Kurzweil, as well as the work of such organizations as
Humanity+, Singularity University, and the Future of Humanity Institute.

What transhumanists of varying stripes hold in common is the belief
that reason, science, and technology can be used to overcome the physical
and cognitive limitations that are endemic to the human condition. Consequently, they are dedicated to "developing and making widely available
technologies to eliminate aging and to greatly enhance human intellectual,
physical, and psychological capacities."[5] In order to pursue such enhance-

4. For examples of some key articles and essays, see Max More and Natasha Vita-More,
eds., *The Transhumanist Reader: Critical and Contemporary Essays on the Science, Technology,
and Philosophy of the Human Future* (Malden, MA, and Oxford: Wiley-Blackwell, 2013).
For critical assessments of the posthuman project, see Francis Fukuyama, *Our Posthuman
Future: Consequences of the Biotechnology Revolution* (New York: Farrar, Straus & Giroux,
2002), and Brent Waters, *From Human to Posthuman: Christian Theology and Technology in
a Postmodern World* (Aldershot, UK: Ashgate, 2006).

5. Max More, "The Philosophy of Transhumanism," in More and Vita-More, eds., *The
Transhumanist Reader*, 3.

ment it must be presumed that there is nothing sacrosanct or given about human nature. Once this pliability is acknowledged, then a highly worthwhile goal can be undertaken to create posthuman beings that "would no longer suffer from disease, aging, and inevitable death."[6]

The holy grail of the posthuman project is personal immortality. Three interrelated strategies are envisioned.[7] The first strategy is to achieve *biological immortality*. With anticipated developments in genetic and biotechnologies, the average lifespan can be increased dramatically, perhaps indefinitely. In principle, there is no reason why the badly concocted DNA bequeathed to humankind through natural selection cannot be reengineered to achieve longer, healthier, and more productive lives.[8]

If human biology proves less malleable than hoped, however, then the second strategy of *bionic immortality* can also be pursued. Advances in prosthetics and brain-machine interface have led to promising therapeutic applications. Mobility, dexterity, and sight, for example, have been restored, and artificial tissue and blood vessels have been manufactured. Again, in principle, there is no reason why with further technological development these therapies could not be used to enhance human well-being. The advantage of this strategy is that not only is longevity increased, but physical and cognitive performance can be improved.

This strategy, however, is not without its risks. Prosthetic limbs, artificial organs, and brain implants, for instance, can malfunction, and a hybrid or cyborg body remains subject to fatal accidents or malicious acts. Although a largely artificial body is an upgrade over its natural counterpart, it is still not an ideal solution in overcoming finite and mortal limits. This limitation prompts the third, and most speculative, strategy: *virtual immortality*. According to such visionary leaders in the fields of artificial intelligence and robotics as Ray Kurzweil and Hans Moravec, the information stored in the brain that constitutes a person's memories, experience, and personality can be digitized. In the near future sophisticated imaging devices will scan the

6. More, "The Philosophy of Transhumanism," 4.

7. For a more detailed account of these strategies, see Brent Waters, "Whose Salvation? Which Eschatology? Transhumanism and Christianity as Contending Salvific Religions," in *Transhumanism and Transcendence: Christian Hope in an Age of Technological Enhancement*, ed. Ronald Cole-Turner (Washington, DC: Georgetown University Press, 2011).

8. See Aubrey de Grey, ed., "Strategies for Engineered Negligible Senescence: Why Genuine Control of Aging May Be Foreseeable," *Annals of the New York Academy of Science* 1019 (June 2004), and "The War on Aging," in Immortality Institute, *The Scientific Conquest of Death* (Buenos Aires: LibrosEnRed, 2004), 29–45.

brain to collect this information and upload it into a computer. Once this information is organized and stored it can be downloaded into a robotic or virtual reality host. With frequently updated and multiple backups, the uploading and downloading process can presumably be repeated indefinitely. Consequently, one's virtual self is virtually immortal. Moreover, one's identity is no longer limited to a singular body, confined by constraints of time and physical location. The information constituting the self can be dispersed simultaneously in various physical and virtual locales, and subsequent experience gained incorporated into oneself or multiple selves. William Sims Bainbridge asserts that multiple personalities may soon become the preferred mode of being. "Buckminster Fuller used to say, 'I seem to be a verb.' Perhaps today we should say, 'I am a plural verb in future tense.'"[9]

Some may protest that a person cannot be reduced to a series of 0s and 1s that can be trundled back and forth between a computer and robotic or virtual reality hosts. But Kurzweil and Moravec reply that since the mind is not a material object, and the mind is ultimately what a person is, then a person cannot ultimately be anything other than information. A personality is comprised of a pattern of organized data created and stored over time. A biological body is merely a natural prosthetic preserving this pattern. Unfortunately, nature has not produced a very dependable or durable prosthetic, so technology must be used to produce a better one. In liberating the mind from the body, the essential information constituting one's identity is not lost. In Moravec's words, "I am preserved. The rest is mere jelly."[10] In short, technology can be developed to save individuals from the poor jelly-like conditions of being human.

It is tempting to dismiss posthuman and transhuman discourse as unbridled speculation of individuals who confuse science with science fiction, or worse, camouflage idle fantasy with a veneer of high-tech jargon. Much of the technological development they envision is highly speculative and in many instances may prove infeasible. Why waste time, then, worrying and arguing about a so-called posthuman future that will probably never come into being? I suspect that this may be true, but the temptation to summarily dismiss posthuman or transhuman discourse should nonetheless be resisted. The danger of posthumanism or tranhumanism is not confined to techno-

9. William Sims Bainbridge, "Transavatars," in More and Vita-More, eds., *The Transhumanist Reader*, 91.

10. Hans Moravec, *Mind Children: The Future of Human and Robot Intelligence* (Cambridge, MA: Harvard University Press, 1988), and *Robot: Mere Machines to Transcendent Mind* (Oxford and New York: Oxford University Press, 1999), 117.

logical feasibility. Indeed, debates over technological feasibility often serve to distract attention away from a far more important issue, namely, that posthuman or transhuman discourse is already shaping the identity, values, and aspirations of late modernity. To a large extent, late moderns already conceive themselves in posthuman terms. In the words of Katherine Hayles: "People become posthuman because they think they are posthuman."[11] Whether the envisioned technological developments prove feasible or not, a posthuman self-perception is already shaping the desires and expectations of late moderns, as well as the actions they take to fulfill them.

In this respect, posthuman or transhuman discourse is not so much predictive and futuristic, but is more akin to hyperbolic description and commentary on late modernity.[12] A so-called posthuman future amplifies to an intense level the desires, hopes, and dreams that late moderns already hold and which they believe can be fulfilled through technological progress. In brief, they believe that nature and human nature can be reduced to its underlying information, and with the right technologies reshaped in more desirable ways. There is no given order other than what is imposed by those with the power to manipulate the pertinent information. Even the human body is perceived as a biological information network that can be reengineered to give humans something better than the "mere jelly" bequeathed to them by evolution.

Since posthuman and transhuman discourse is both depictive and interpretive of present circumstances, then it may also be characterized as an initial attempt at creating a myth. A myth narrates origin and destiny, and how good contends with evil in between, eventually overcoming it in the end. A myth, then, is not a fairy tale or fable, but a narrative ascription of the human condition; a literary device that encapsulates where hope and trust should be placed, in turn aligning desires accordingly. Humans cannot live without their myths, for as Jonathan Gottschall argues, they are hopelessly storytelling creatures.[13] Some myths are more compelling than others, because they offer what is believed to be a more truthful interpretive account of the human condition. C. S. Lewis's conversion to

11. N. Katherine Hayles, *How We Became Posthuman: Virtual Bodies in Cybernetics, Literature, and Informatics* (Chicago and London: University of Chicago Press, 1999), 6.

12. For a more detailed and critical analysis of this hyperbole, see Brent Waters, *Christian Moral Theology in the Emerging Technoculture: From Posthuman Back to Human* (Farnham, UK, and Burlington, VT: Ashgate, 2014).

13. Jonathan Gottschall, *The Storytelling Animal: How Stories Make Us Human* (New York: Houghton Mifflin Harcourt, 2012).

Christianity, for example, was prompted in large part by the attraction of its mythology.

Moreover, the posthuman myth is a salvific myth, for humans must be saved from their finitude and mortality. In this emerging narrative, nature displaces the Fall as the challenge to be overcome. Humans are not suffering the degenerative consequences of a lost pristine condition. Rather, nature is the culprit because through its clumsy selection process it prevents humankind from fully developing its latent potential. Consequently, nature itself must be domesticated and redirected so that humans can be all that they can be—and more. As Max More, a leading transhumanist philosopher, contends in his "Letter to Mother Nature":

> Mother Nature, truly we are grateful for what you made us. No doubt you did the best you could. However, with all due respect, we must say that you have in many ways done a poor job with the human constitution. You have made us vulnerable to disease and damage. You compel us to age and die—just as we are beginning to attain wisdom. . . . What you have made is glorious, yet deeply flawed. . . . We have decided that it is time to amend the human constitution.[14]

Accomplishing this amendment requires reorienting human evolution toward the end of becoming posthuman, entailing the enhancement and eventual perfection of physical and cognitive capabilities through technological intervention.

It is admittedly odd to accuse posthumanists and transhumanists of being mythmakers given the strong religious connotation. The most prominent pundits take great pride in portraying posthumanism and transhumanism as a movement guided by "reason, technology," and the "scientific method" instead of any religious belief or hope in "supernatural forces."[15] More concedes that some religious beliefs can purportedly be reconciled with transhumanism, and there are a few Christians, Mormons, Buddhists, and Jews that also identify themselves as transhumanists. But this is "very rare," and such beliefs are idiosyncratic, exerting little, if any, influence on the posthuman movement.[16] More baldly, Russell Blackford asserts: "Most typi-

14. As quoted in Christina Bieber Lake, *Prophets of the Posthuman: American Fiction, Biotechnology, and the Ethics of Personhood* (Notre Dame: University of Notre Dame Press, 2013), 95.

15. More, "The Philosophy of Transhumanism," 4.

16. More, "The Philosophy of Transhumanism," 8.

cally, transhumanists embrace a naturalistic and purely secular worldview. In short, transhumanism is not a religion."[17] Hope for a posthuman future is thereby based on "human creativity rather than faith."[18] In addition, posthumanists and transhumanists would object that they are not on a quest for perfection—they are not utopians but extropians dedicated to a perpetual and unending process of change and improvement.[19] Consequently, are they not creating a myth of an imperfect (dare we say fallen?) nature being re-engineered (dare we say saved?) toward the goal of an immortal (dare we say eschatological?) posthuman future?

This is a curious defense for claiming that one is innocent of mythmaking. Posthumanism and transhumanism are not religions centered around formal liturgical and spiritual practices, but their rhetoric is thoroughly laced with statements of faith. Following Martin Luther, what one's heart clings to is effectively one's god, and posthumanists and transhumanists do a lot of clinging to human creativity, particularly when it is armed with technological capability. To place one's trust and hope in human creativity is a leap of faith and not solely the outcome of reason or the scientific method. The claim that no quest for perfection is being undertaken is strange on both epistemological and practical grounds. How can we know if progress or improvement is perpetual or endless? Moreover, how can we determine if we are actually progressing or improving if there is no end of perfection against which certain actions may be measured? And even if progress is perpetual, is not the goal of fully embracing endless improvement, effectively achieving a state of perfection? To claim that the human condition can be improved by replacing natural selection with human creativity is a statement of faith that in turn drives a mythological imagination.

Prominent narrations of how a posthuman future will be achieved and what it will entail are almost always propounded in mythical terms. Two examples may suffice to illustrate. According to Ray Kurzweil, the Singularity is near.[20] With anticipated developments in IT, AI, and nanotechnology the "pace of technological change will be so rapid, its impact so deep, that human life will be irreversibly transformed."[21] Kurzweil's prediction is based

17. Russell Blackford, "The Great Transition: Ideas and Anxieties," in More and Vita-More, eds., *The Transhumanist Reader*, 421.

18. More, "The Philosophy of Transhumanism," 4.

19. See More, "The Philosophy of Transhumanism," 14.

20. See Ray Kurzweil, *The Singularity Is Near: When Humans Transcend Biology* (New York and London: Penguin Books, 2005).

21. Kurzweil, *The Singularity Is Near*, 7.

on his belief that evolution increases order. Technological development is a more efficient evolutionary process, and humans can use its superior order to transcend the less efficient constraints of biological evolution. By constructing a superior computational substrate, the human mind will grow larger and more diffuse. Since computation underlies everything that is important, human intelligence will dramatically and rapidly increase by using an artificial computational foundation.[22] Moreover, Kurzweil insists "that within several decades information-based technologies will encompass all human knowledge and proficiency, ultimately including the pattern-recognition powers, problem-solving skills, and emotional and moral intelligence of the human brain itself."[23] Since posthumans will essentially be minds hosted by a variety of artificial bodies and substrata they will exert greater control over their own fate, enabling them to live for as long as they desire. The Singularity entails the total merger of human biology and technology. More broadly, there "will be no distinction, post-Singularity, between human and machine or between physical and virtual reality."[24] Most importantly, the emergence of the Singularity will be rapid, steadily progressive, and inevitable. Invoking Moore's law on the exponential growth of computational speed, Kurzweil extrapolates that the Singularity will emerge sometime around 2045, barring any unforeseen natural or humanly produced global catastrophes.

Even among posthumanists and transhumanists Kurzweil is regarded as being a bit too ambitious and optimistic. Although a posthuman future is humankind's proper destiny, it may come about more haltingly and later than he envisions. Ted Chu, for example, believes that a posthuman future is humankind's destiny, but the way forward will be fraught with egregious twists and retrograde steps.[25] Rather than turning to Gordon Moore, Chu, an economist, is more influenced by Joseph Schumpeter's concept of creative destruction as a lens for understanding the evolutionary process.[26] Natural evolution is a clumsy, if not brutal process, entailing as many, if not more, failures and dead-ends as successes that are also often short-lived. It is this

22. See Kurzweil, *The Singularity Is Near*, 127–28.

23. Kurzweil, *The Singularity Is Near*, 8.

24. Kurzweil, *The Singularity Is Near*, 9.

25. Ted Chu, *Human Purpose and Transhuman Potential: A Cosmic Vision for Our Future Evolution* (San Rafael, CA: Origin, 2014).

26. See Joseph A. Schumpeter, *Capitalism, Socialism and Democracy* (New York and London: Harper Perennial, 2008), esp. chap. 7.

destructive dynamism, however, that poses both the challenge and hope for constructing a posthuman future.

The challenge is for humans to transcend the constraints of their natural or biological evolution. Humans must be willing to "fight for transcendental freedom from the genetic tyranny that natural history has imposed on us."[27] The hope is that humans can achieve this transcendence by directing their cultural evolution toward a posthuman end.[28] This redirection requires individuals to forgo their short-term self-interests in order to contribute to the far greater purpose of cosmic evolution. The "effort to consciously participate in the unfolding of this grand evolutionary process is the greatest purpose we can find and identify with, and it constitutes a new kind of heroism for our time."[29] Humankind has reached a "threshold" at which "modernity and science seem to be putting the human species in danger, but at the same time they are offering the unprecedented, seemingly incredible promise of a quantum leap in the human condition."[30]

Chu insists that navigating this threshold in a positive manner requires a spiritual renewal. He contends that retrieving ancient wisdom will help humans to choose wisely in redirecting their cultural evolution toward a posthuman future. He is also not embarrassed to invoke religious insights, although he draws selectively upon an eclectic cohort of writers, and is particularly fond of Eastern sages and heterodox Christians. These ancients had an inkling that the universe was not about humans, but they lacked the scientific knowledge and technological capability to reorient their lives appropriately. Late moderns do not suffer this deficiency, and if they can muster the courage to admit that the "human era as we have known it is coming to an end," and the "posthuman era is about to begin,"[31] then they can redirect their cultural evolution accordingly. The challenge, then, is for humans to embrace this "calling" to serve the greater purpose of cosmic evolution. This calling, however, is not based on a "leap of faith" but following the trajectory of human evolution as a species, despite its many twists and turns.[32] In embracing this calling, humans discover that their true "purpose is to transcend their limiting biology," thereby "enabling the rise of new kinds of sentient beings, freed from our genetic limitations in

27. Chu, *Human Purpose and Transhuman Potential*, 10.
28. See Chu, *Human Purpose and Transhuman Potential*, 21.
29. Chu, *Human Purpose and Transhuman Potential*, 9.
30. Chu, *Human Purpose and Transhuman Potential*, 8.
31. Chu, *Human Purpose and Transhuman Potential*, Kindle reader location 206.
32. Chu, *Human Purpose and Transhuman Potential*, loc. 200.

pursuit of the highest transcendental aspirations and the promotion of cosmic evolution."[33] In attempting to transcend their biological limitations, humans both acknowledge the challenge and embrace the hope entailed in this transcendence, for the "posthuman future is not *about* us per se, but it is *up to* us to make it happen. Humanity as an end in itself is hopeless, yet all hope for the future resides in humanity."[34] Humans, particularly late moderns, carry a heavy burden, for they must will their eventual extinction in order to assist the emergence of a superior posthuman being; it is a quintessential act of creative destruction.

Although Chu offers a more restrained vision than that of Kurzweil, both share a confident optimism that the destiny of humankind is a posthuman future, and in both instances it is a future largely devoid of humanity in any recognizable form.[35] In this respect, both are narrating an emerging posthuman myth, albeit in different ways, that a truculent nature must and shall be replaced by a more accommodating artifice, that an indifferent natural evolution must and shall be displaced by a purposeful cultural evolution. And in both accounts transformative technology is the means of salvation.

What are we to make of this myth? I offer a few brief, principally interrogative, and at times cryptic observations.

The first observation involves intellectual lineage. Almost without exception, posthumanists and transhumanists proudly proclaim themselves to be children of the Enlightenment. Many characterize themselves as hypermodernists, at odds with, if not battling, both postmodernists and so-called bioconservatives; hence, the emphasis on reason, freedom, individual autonomy, and progress. Consequently, it is reasonable to allow individuals to enhance themselves in whatever ways they might please so long as they do not harm others in the process. Moreover, the quest to become posthuman continues the Enlightenment's belief in progress and the ability to use reason creatively in bettering the human condition.

Like most lineages, there are also some embarrassing ancestors best consigned to the attic but who nonetheless exert a formative influence on

---

33. Chu, *Human Purpose and Transhuman Potential*, loc. 194.

34. Chu, *Human Purpose and Transhuman Potential*, 20 (emphasis original).

35. It should be noted that Bill Joy, co-founder of Sun Microsystems, is now largely regarded as a renegade or infidel by posthumanist and transhumanist cabals in response to his article, "Why the Future Doesn't Need Us," that was published in *Wired* magazine (April 2000). He argues that many of the technologies championed by the likes of Kurzweil, Chu, More (and that Joy helped invent) will not usher in a new posthuman age but will more likely result in human extinction.

their descendants.[36] In the case of posthumanists and transhumanists, two such ancestors include a theological and scientific heretic. The theological heretic is Pelagius. For Pelagians, humans become what they will themselves to be. To become perfect you must will yourself to be perfect. If you lack the ability to understand what perfection entails or your will proves weak, then you must be prepared to suffer the ill consequences. Pelagians, as I argued above, align their will with the good in order to achieve perfection. Posthumanist Pelagians align their will with the Singularity or cosmic evolution to achieve perfection in this respect: humans must choose either to align their wills with the good of Kurzweil's Singularity or Chu's cosmic evolution, or to will the evil of an incompetent Mother Nature. In the hands of latter-day Pelagians, such as posthumanists and transhumanists, there is also the caveat that the slow-witted and weak-willed should not be allowed to prevent the enlightened and strong-willed from pursuing the good of their perfection (or perpetual improvement if you prefer). The scientific heretic is Lamarck. Curiously, posthumanists and transhumanists love evolution but despise natural selection. This is why culture must displace nature as the principal force driving human evolution toward a posthuman future, for unlike biological reproduction, acquired characteristics can be passed on from one generation to the next. The power of culture to direct evolution becomes all the more pronounced if extensive germline and/or bionic modification prove feasible.[37]

Children should not be held accountable for the sins of their ancestors, but descent cannot be entirely dismissed. There are a number of troubling strands in the posthuman and transhuman intellectual lineage that again can only be briefly noted. It is far from apparent that the Enlightenment project is an unqualified good. As the voluminous literature of its critics and defenders demonstrate, it is at best a mixed bag. But if the posthumanists and transhumanists are hypermodernists, then is not the envisioned posthuman future little more than the Enlightenment project writ large (very large)

36. It is interesting to note that in the forthcoming second edition of the *Encyclopedia of Ethics, Science, Technology and Engineering* the authors of the entry on "Transhumanism" claim that its three principal intellectual strands are the Enlightenment, Pelagianism, and Arianism.

37. Lamarck's work influenced a number of nineteenth-century Protestant theologians. Horace Bushnell, for example, argued that if sin can be passed down from parent to child, then so too can piety or sanctity. Consequently, the best way to evangelize the world is for a "sanctified stock" of Christians to outbreed inferior competitors. See Horace Bushnell, *Christian Nurture* (New Haven: Yale University Press, 1947), esp. chap. 8.

complete with its wild gyrations between unprecedented humanitarian concern and unspeakable atrocities, tendencies that may very well be amplified with increased technological capabilities?[38] Cannot the Enlightenment's late modern children be characterized, to a limited extent, as killer angels; as beings that have honed their skills for achieving both humane accomplishments *and* committing unspeakable atrocities? How else can we account for the simultaneous acts of compassion *and* cruelty that plague late modernity? And is there is any compelling reason to believe that *both* these capacities will be amplified through greater technological capability? This is why Katherine Hayles worries that what is "lethal is not the posthuman as such but the grafting of the posthuman onto a liberal humanist view of the self."[39]

Pelagianism is a one-trick pony in which the only ethical counsel on offer is ultimately to try harder. This burden can perhaps be borne by the enlightened and the strong-willed, but it crushes all others, a necessary culling if a posthuman future is to be achieved. Despite the liberal and humanistic rhetoric of many posthumanist and transhumanist writers, they are engaged in a thoroughly elitist enterprise, and elites, more often than not, grow weary of the distracting fears and anxieties imposed by the lowly and unfit. Whether nature or culture proves to be the better selector remains to be seen. There may be some good reasons for placing faith and hope in the latter, but this does not warrant any certainty that a posthuman future will be a golden age populated with angel-like beings. As Chu reminds his posthumanist and transhumanist colleagues, even cultural evolution is more Darwinian than they often care to admit.

The second observation involves Passmore's contention that all quests for perfection inevitably fail. Will the posthuman project escape this pattern? Perhaps, but I think the odds are against it. The problem is that once humans convince themselves that they know, with a great deal of certainty, the difference between good and evil, a great deal of mischief results. When a course of action is undertaken to achieve the good, a peculiar logic emerges on what to do with those standing in the way, often justifying recourse to coercion. The posthumanists and transhumanists are fairly certain that it is good to become posthuman and evil to remain merely human. Despite their rhetoric of toleration for those not sharing their vision of the future—no one will be forced to enhance himself or herself—how long can they tolerate evil

38. See Colin Gunton, *The One, the Three, and the Many: God, Creation, and the Culture of Modernity* (Cambridge: Cambridge University Press, 1993).
39. Hayles, *How We Became Posthuman*, 286–87.

to prevent this good transformation? It is not hard to imagine that over time individuals refusing to extend their longevity, and enhance their physical and cognitive performance, or withhold these benefits from their offspring, will be stigmatized or perhaps even penalized as being irresponsible or worse. As Hannah Arendt observed, the French Revolution was prompted by pity for the poor and destitute, and ended in an orgy of executions and massacres of those standing in the way of the new age of liberty, equality, and fraternity.[40]

The final observation involves how Christians should respond to this posthuman myth. In many respects, the posthuman myth is a bad and distorted retelling of the Christian myth. A fallen creation is exchanged for an indifferent and inefficient nature. Salvation by the triune God through the person of Jesus Christ is substituted by a triumvirate of human reason, creativity, and technological development. An eschatological hope for eternal fellowship with God is replaced by a hope in the immortal posthuman. The inadequacy of these substitutions can be easily demonstrated, but what is most troubling in this eviscerated myth is what is missing: there is no incarnation, and the narrative is devoid of grace and forgiveness.

For posthumanists and transhumanists, whatever transcendence humans might encounter is the result of their own efforts, for there is no divine being that may enter the human condition. Consequently, the task is to transform flesh into data rather than encountering the Word made flesh. With this reversal it is understandable why the body is regarded as little more than mere jelly, for it is an unwanted burden that constrains the immaterial will. Embodiment, in short, is an evil to be overcome. Posthumanists and transhumanists might reply that they do not loathe their bodies. As More contends, the body is a "flawed piece of engineering," delimiting "morphological freedom," and rather than "denying the body transhumanists typically want to choose its form and be able to inhabit different bodies, including virtual bodies."[41] If this is a purported affirmation of the body, one shudders to think what a denunciation entails.

What the incarnation affirms, in part, is the creaturely status of the human, a status that is inherently and necessarily finite and mortal.[42] Since the Word was pleased to be born and dwell among humans, to take on their status, then humans may in turn affirm the goods of finitude and mortal-

---

40. See Hannah Arendt, *On Revolution* (New York: Viking, 1965).

41. More, "The Philosophy of Transhumanism," 15.

42. For a more extensive account of finitude and mortality as human goods, see Brent Waters, *This Mortal Flesh: Incarnation and Bioethics* (Grand Rapids: Brazos, 2009).

ity. In the incarnation, God embraces humans for what they are and not what they might prefer to be. This embrace, however, does not imply that nothing should be done to ameliorate suffering or improve human life and lives. There is no reason why Christians should not endorse or participate in developing medical and technological advances that enhance healthcare or provide greater comfort. Improving physical capabilities, however, is not synonymous with attempting to supplant the finitude and mortality of purportedly imperfect bodies. It is one thing, for instance, to boost the immune system to resist disease, and quite another to transform humans into immortal beings that are impervious to disease. Rather, the incarnation reminds humans of their fallen state that should in turn prompt a way of life that is not oriented toward eliminating that which is mistakenly perceived as imperfection, resulting in ill-fated quests for perfection. In affirming what the incarnation affirms, humans also confirm the bonds of imperfection that bind them together as fallen creatures.[43]

It is because of and in response to these imperfect bonds of fellowship that there is a need for forgiveness and grace. As fallen creatures—sinners, to use a terribly unfashionable term—humans harm each other through both acts of commission and omission. The giving and receiving of forgiveness is required to restore ruptured fellowship; a life that is genuinely human is habitually forgiving. It is in the church that Christians learn, in part, how to offer and receive forgiveness, and are in turn formed as habitually forgiving and forgiven people. The Eucharist, for example, encapsulates this formative practice. The Eucharistic liturgy begins with judgment, confession, and repentance and ends with a promise of amendment of life before absolution is pronounced. The Lord's Table reminds believers that as fallen creatures they sin often against God and neighbor, but the forgiveness offered does not excuse their sins, but through grace discloses and judges them. Following C. S. Lewis in his essay "On Forgiveness," to accept forgiveness is to acknowledge responsibility for one's blameful act, whereas an excuse entails extenuating circumstances for which one cannot be blamed.[44] A good excuse means there is nothing to forgive, or, borrowing from Hannah Arendt, only that which can be rightfully punished can be forgiven.[45] Hence absolution cannot be given until culpability is confessed and the promise of amending

43. See Oliver O'Donovan and Joan Lockwood O'Donovan, *Bonds of Imperfection: Christian Politics, Past and Present* (Grand Rapids: Eerdmans, 2004).

44. See C. S. Lewis, *The Weight of Glory and Other Addresses* (HarperCollins e-books, 2009), 177–79.

45. See Hannah Arendt, *The Promise of Politics* (New York: Schocken Books, 2005).

one's life accordingly is made, a promise that is frequently broken by fallen creatures, initiating subsequent cycles of confession and repentance.

Presumably, posthumanists and transhumanists have nothing to say about forgiveness because in transforming humans into posthumans they are doing nothing wrong that needs to be forgiven. For why would a quest for perfection (or endless improvement if you prefer) be regarded as an act to be punished, and thereby forgiven? Yet this is precisely the trap Passmore identifies, namely that those seeking perfection pursue what they believe is good but invariably settle for power, be it over nature, themselves, or others. In the current quest for an immortal posthuman future, ancient Greek mythology provides a sobering warning: the gods were both powerful and immortal, but they were rarely lifted up as paragons of virtue, goodness, much less perfection. Or to change the mythology, are the posthumanists endeavoring to construct a new Eden that has a forest of bountiful trees of knowledge, particularly of the Baconian variety that also bring power? But is there any tree of life to be found in this new garden?

Christians engage the late modern world as forgiven people, and as such are reminded of their fallen state. This reminder helps to account for the cautionary note they should strike in assessing all acts for improving the human condition, for like themselves they encounter other fallen humans who know good and evil but lack the wisdom to discern fully their difference. Particularly in respect to the latest attempt to perfect humans by making them posthuman, the words of Dietrich Bonhoeffer are instructive: "The knowledge of good and evil seems to be the aim of all ethical reflection. The first task of Christian ethics is to invalidate this knowledge."[46]

---

46. Dietrich Bonhoeffer, *Ethics* (London: SCM, 1955), 3.

8 On Learning to See a Fallen and Flourishing Creation

....

Alternate Ways of Looking at the World

NORMAN WIRZBA

To "look" and to "see" are vastly different things. Though people may look at the same scene, what individuals see can vary considerably. This is because every viewer comes equipped with different perceptive faculties or habits of attention, and with varying desires, fears, questions, and agendas. To look inevitably presupposes a perspective or point of view that is itself a reflection of one's physical *location*, one's *time*, one's philosophical and religious *commitments*, and one's *standing* within a culture.

Though *looking* may presuppose little more than the sensory capacity for sight, *seeing* presupposes what Hans-Georg Gadamer called a "hermeneutical consciousness." To see is to interpret, and to interpret is to put to practical use languages, concepts, and symbolic systems of varying kinds that enable us to sense the meaning of what we look at. To see, in other words, is to *understand* in particular sorts of ways what one is perceiving.[1]

---

1. Though I employ sight in this chapter as the metaphor for understanding the world, it should be clear that other senses, like touch and smell and taste, should not be ignored, particularly since they often lead to a more embodied, practical, and intimate relationship with the world. In *Food and Faith: A Theology of Eating* (New York: Cambridge University Press, 2011) I argue that taste, along with the embodied practices of food's production and consumption, open fresh lines of inquiry and sympathy as we move to understand where we are. To this can be added the important new collection of essays *Carnal Hermeneutics*, ed. Richard Kearney and Brian Treanor (New York: Fordham University Press, 2015), on the body as a site of interpretation. The hegemony of sight in philosophical traditions of inquiry, and the distancing of self and world it often presupposes, is well described by Martin Jay in *Downcast Eyes: The Denigration of Vision in Twentieth-Century French Thought* (Berkeley:

This means that seeing the world *as* "fallen" is not obvious, nor is it straight-forwardly scientific. To see fallenness people must have in place a theologi-cally inspired interpretive framework or hermeneutical consciousness that allows them to judge the world and its creatures in a particular sort of way. Alexander Schmemann puts it this way:

> The world is a fallen world because it has fallen away from the awareness that God is all in all. The accumulation of this disregard for God is the original sin that blights the world. And even the religion of this fallen world cannot heal or redeem it, for it has accepted the reduction of God to an area called the "sacred" ("spiritual," "supernatural")—as opposed to the world as "profane." It has accepted the all-embracing secularism which attempts to steal the world away from God.[2]

The discipline of hermeneutics teaches that there is no unmediated encounter with the world because to be in a world is always already to be engaged in acts of interpretation that "open" the world as a place that can be understood and engaged. Drawing on the work of his teacher Martin Heidegger, Gadamer argued that "understanding is not just one of the various possible behaviors of the subject but the mode of being of Dasein itself." Hermeneutics "denotes the basic being-in-motion of Dasein that constitutes its finitude and historicity, and hence embraces the whole of its experience of the world."[3] None of us exist in a neutral space. None of us simply "look" at things, having no interest at all in what is observed (a completely disinterested looking would be like an unfocused camera lens that produces no discernible image at all). From the moment we are born we are being educated, whether formally or not, to see, to focus, to evaluate, and thus also to engage our surroundings in the unique ways that we do.

This chapter is my exploration of one compelling theological framework

---

University of California Press, 1994) and the collection *Modernity and the Hegemony of Vision*, ed. David Michael Levin (Berkeley: University of California Press, 1993).

2. Alexander Schmemann, *For the Life of the World* (Crestwood, NY: St. Vladimir's Seminary Press, 1963), 16.

3. Hans-Georg Gadamer, *Truth and Method*, 2nd ed. (New York: Crossroad, 1991), xxx. For a wide-ranging discussion of the implications of hermeneutics for our understanding of the natural world see *Interpreting Nature: The Emerging Field of Environmental Hermeneutics*, ed. Forrest Clingerman, Brian Treanor, Martin Drenthen, and David Utsler (New York: Fordham University Press, 2014).

in which creation's *fallenness*, but also its *flourishing*, becomes intelligible. In particular it is an examination of how a Christological understanding of creation makes possible an understanding of the world as fallen. Fallenness is not some general feature of the world that can be seen by just anyone. It is a corollary of the action of sin in the world. Its resonance or meaning comes from being with Christ as the one who helps us see the world as the place of God's love, but also as the place that has been wounded by love's deformation. To say that the world is fallen is to say that the love of God is not freely and fully active within it. It is to say that the world still awaits its fulfillment and perfection when God's love will be all in all. As I will suggest, Jesus stands with us and helps us bring the world into focus in certain sorts of ways. He tells us how and where to look, lets us know when we are out of focus, and equips us to see the significance of what is going on around us. This means that Christian discipleship is not only humanity's entrance into life with God. It is also humanity's introduction to the world now understood and engaged in a new way.

Interpreting the World

How people have understood the world as a whole has varied greatly through time. Sticking simply to the ancient Greek philosophical context, when Democritus looked at the world he "saw" invisible, indivisible *atomoi* in perpetual, random motion. There is no force or intelligence directing their coming to be or their falling apart. Stuff simply happens! This picture of an atomist, pluralist world was in striking contrast to that of Anaxagoras, who believed that the various elements of the world share in each other and in the whole. Moreover, there is nothing that is accidental or random about this world because *Nous* or Mind permeates the whole, giving it the shape and form that it does. For Anaxagoras the world forms an ordered, intelligible whole, a *kosmos*.

Why these dramatically different ways of seeing the world? Is it that Anaxagoras's picture (potentially) yields a more rational, regular, and reliable world in which people can say that whatever happens happens for a reason or perhaps as a witness to Fate? Or is it that a picture of the world is in some sense also a picture of ourselves, reflecting what we hope to see in the world?

Following Pierre Hadot, it is important to underscore that ancient philosophy, and the "science" it made possible, was first and foremost about the advocacy for a way of life and the disciplines that enabled its practitioners to live well (however that was conceived). *Theōria*, the way of seeing being recommended by a philosophical school, was inextricably connected to an *ethos* or way of being in the world. To the extent that one's picture of the world did not serve to help people live better lives, one ceased being genuinely philosophical.[4] The whole point in serious contemplation of the world was to effect self-transformation, which meant that an *ethos* was accompanied by an *askēsis*, a form of asceticism or personal discipline that aligned the life of the wisdom seeker with the truth of the world. *Theōria, ethos*, and *askēsis* are inextricably intertwined. As we will see, their interconnection is clearly at work in early Christianity as well: seeing the world in a Christian manner, and thus being able to judge it as fallen or flourishing, were intertwined with living in the world in particular sorts of ways.

It is clear that this ancient manner of characterizing philosophical reflection has not been universally upheld. Though more contemporary philosophical pictures of the world may not immediately or obviously be in service of what is believed to be a better *askēsis* or way of life, it is nonetheless apparent that people are encouraged to see and understand the world in ways that serve some interest or goal, even if that goal is not explicitly stated or reflected upon. *What* people are asked to see, the *modes* and the *tools* they are given to look at it, the *categories* and frames they are provided to organize what they see, and the *significance* they are supposed to discern as a result of their looking—all these are more or less established before and while they take their various looks. Hadot observes:

> University philosophy therefore remains in the same position it occupied in the Middle Ages: it is still a servant, sometimes of theology, sometimes of science. In any case, it always serves the imperatives of the overall organization of education, or, in the contemporary period, of scientific research. The choice of professors, course topics, and exams is always subject to "objective" criteria which are political or financial, and unfortunately all too often foreign to philosophy.[5]

4. Pierre Hadot, *What Is Ancient Philosophy?* (Cambridge, MA: Harvard University Press, 2002), 172–233. Put succinctly, "in antiquity it was the philosopher's choice of a way of life which conditioned and determined the fundamental tendencies of his philosophical discourse" (272–73).

5. Hadot, *What Is Ancient Philosophy?*, 260.

What we see, in other words, is a feature of the institutions, professional protocols, personalities, streams of financial funding, and packaging of data that open whatever point of view we happen to occupy.[6]

My point is not to say that we see whatever we want. It is, rather, to note that the *theōria* that enables us to make sense of what we are looking at always develops within an *ethos* and an *askēsis* that opens, directs, and disciplines our access to the world. To appreciate what I mean it is helpful to look at the process of seeing as it happened in the work of Charles Darwin.

In his autobiography Darwin tells us that the "gloomy parson" Thomas Robert Malthus's essay on population played a decisive role in his own work because it gave him the categories that enabled what he looked at to come into a more compelling focus:

> Fifteen months after I had begun my systematic inquiry, I happened to read for amusement Malthus on Population, and being well prepared to appreciate the struggle for existence which everywhere goes on, from long continued observation of the habits of animals and plants, it at once struck me that under these circumstances favorable variations would tend to be preserved and unfavorable ones to be destroyed. The result of this would be the formation of new species. Here I had at last got a theory by which to work.[7]

Here we can observe how Malthus gave Darwin the optics or interpretive lens by which to see the world as signifying in particular sorts of ways. Darwin had been looking at the world for a long while, but he had not yet found the interpretive framework that enabled him to make satisfactory sense of what he was looking at. Malthus gave him the hermeneutical framework he longed for. His *theōria* enabled Darwin to see things of all sorts as waged in competitive struggle and war so that they might increase themselves in face of scarce and diminishing resources. As Darwin would write in *The Descent*

6. For a rigorous and wide-ranging examination of the various modes whereby truth conditions are established and legitimated, see Bruno Latour's *An Inquiry into Modes of Existence: An Anthropology of the Moderns* (Cambridge, MA: Harvard University Press, 2013). Latour delineates the many values and modalities people have employed to experience and understand "reality," and shows how the scientific, social, and economic framings of "experience" overlap and come apart to make possible the regional ontologies that make our worlds meaningful.

7. Quoted in Conor Cunningham's *Darwin's Pious Idea: Why the Ultra-Darwinists and Creationists Both Get It Wrong* (Grand Rapids: Eerdmans, 2010), 9–10.

of Man, all organic beings expend effort to increase their numbers. These populations, much like the human populations Malthus described, increase geometrically, and in places that cannot keep up with such rapid levels of increase. "Hence, as more individuals are produced than can possibly survive; there must in every case be a struggle for existence, either one individual with another of the same species, or with individuals of distinct species, or with the physical conditions of life. It is the doctrine of Malthus applied with manifold force to the whole animal and vegetable kingdom."[8]

Darwin's seeing of the world is saturated with an *ethos* of scarcity that also reflects an *askēsis* of unremitting struggle and competition. It yields a vision of the world famously described by the poet Tennyson as nature "red in tooth and claw." To look at any organic being is to see a drive to grow and reproduce itself. If such a being is to survive it must adapt to changing circumstances or die because it is only the "fit" beings, those who can best utilize the place they are in to improve reproductive potential, that can thrive.[9]

As a way of seeing the world, a Malthusian/Darwinian *theōria* clearly has considerable explanatory power. A lens focused on the struggle for survival brings multiple elements of the world into focus. Moreover, Darwin's insight into creatures embedded within and in continuity with other creatures is, in my view, essential. It would be naïve, however, to think that his account of the world is "objective" or "comprehensive" in any straightforward meaning of the terms, or that it brings everything into focus. What does his *theōria* leave out of view and out of consideration, and what might it prevent lookers from seeing? Why should we think that species self-interest is the power at work in natural selection, particularly if we begin to unpack the complexity of terms like "self" and "interest" and "select"? Why assume scarcity in a world that might also be characterized by great abundance? Why believe that the drive to live is a drive primarily to "survive" rather than "thrive" or "delight"? These are just some of the questions we can ask about the kind of seeing that follows from a Darwinian framework. This way of seeing, as valuable as it is, is not the only way of seeing. It bears noting that among

8. Darwin, *On the Origin of Species,* quoted in Cunningham, *Darwin's Pious Idea,* 10.

9. It is important to stress that Darwin made room for concepts of cooperation and community in his work, and that more recent evolutionary theory has developed these themes in very important ways; see especially *Evolution, Games, and God: The Principle of Cooperation,* ed. Martin A. Nowak and Sarah Coakley (Cambridge, MA: Harvard University Press, 2013), for an excellent overview and development of these themes. I focus here on the themes of struggle and survival because these are the ones that have most captured the imaginations of laypeople.

indigenous peoples, peoples whose livelihood depended on detailed and careful observation, it is common to find them picturing (and understanding) a world governed by kinship and generosity rather than competition and scarcity. It also bears noting that this Darwinian hermeneutic makes it very difficult to speak meaningfully about fallenness because the idea that the world and its creatures are moving toward a telos or goal in God is impossible to sustain: there is no goal to ecosystem processes or species behavior beyond survival itself, and no real critique of the various means of survival that are possible. As we will soon see, a Christian understanding of creaturely fallenness presupposes that the divine love animating the world has become distorted or denied and that creatures are unable to find their fulfillment in God.

It is, nonetheless, important to consider carefully Darwin's vision of the world because several of his key concepts—fitness, scarcity, survival of the fittest—have made their way into the diverse disciplines of today's education. Darwin is invoked not only to describe what we might call the natural world. He has also given the basic tools by which social worlds are described and explained and (sometimes) justified, which is to say that in Darwinism we now find a philosophical picture or *theōria* of the world that is in service to an *ethos* or particular way of being in the world. Numerous scholars, for instance, have noted that Darwin's picture of ecology is strikingly similar to Adam Smith's picture of economy: both presuppose a vision of things in which individuals operate in ways to maximize self-interest. Both assume processes in which weakness is eliminated to make room for the strong. Both assume a picture of individuals fearful of not getting enough.

It would be a mistake to dismiss Darwin's scientific observations. But it would also be a mistake not to note the narrowness of what might be called its moral vision. Marilynne Robinson, for instance, observes: "That human beings should be thought of as better or worse animals, and human well-being as a product of culling, is a willful exclusion of context, which seems to me to have remained as a stable feature of Darwinist thought. There is a worldview implicit in the theory which is too small and rigid to accommodate anything remotely like the world."[10] What is missing is a world that makes room for the soul.[11] What is missing is a world in which charity—the

10. Marilynne Robinson, "Darwinism," in *The Death of Adam: Essays on Modern Thought* (New York: Picador, 1998), 46–47.

11. Robinson develops this theme in *Absence of Mind: The Dispelling of Inwardness from the Modern Myth of the Self* (New Haven: Yale University Press, 2010). Here Robinson defends the human mind as more than a material mechanism and as having the ability to,

very virtue that would enable us to see and address the misery of the weak—has much force.

This brief look at Darwin helps us see that *theōria*, indeed the whole production of reason, is never innocent. *Theōria* is never far removed from an *ethos*, and that means our looking is invariably in service of or in response to particular concerns, anxieties, ambitions, or desires, i.e., every *theōria* recommends, grows out of, and is in service of an *askēsis* or way of being in the world. Though people might think that their reasoning is clear, logical, persuasive, perhaps even comprehensive, the history of humanity's attempts at reasoning shows that humanity's efforts to clarify the world often have the effect of distorting, dissimulating, even brutalizing it. Idolatrous seeing is an ever-present temptation.[12] It is important to remember that the long march of Western philosophical and scientific development has led to the imperial conquest of the world's continents, the genocide of many of its indigenous populations, and the systematic plundering, pollution, and degradation of the world's habitats. Never before in the history of humanity have we been able to look upon the earth with such precision and breadth. Never before have we witnessed so much degradation that is the result of how we see.

We are, in short, in the midst of a crisis of seeing. The faith once given to philosophers has been transferred to technicians and economists who, it is commonly believed, will present the world "truly," and give the means by which to live conveniently and comfortably within it. But even this faith is wavering as people sense various forms of environmental catastrophe threatening the viability of the very world scientists and technicians are helping create. To be sure, scientists and philosophers have given us a great number of gifts in the forms of engineering, medicine, and education, but it would be naïve to ignore that today's research institutions and machinery are leading us to the extinction of human life.[13] Our doom may not come about in one

---

among other things, reflect morally about the world. We need to be able to affirm that "the strangeness of reality consistently exceeds the expectation of science, and that the assumptions of science, however tried and rational, are very inclined to encourage false expectations" (124).

12. I develop the character of idolatrous seeing in *From Nature to Creation: A Christian Vision for Understanding and Loving Our World* (Grand Rapids: Baker Academic, 2015).

13. I have in mind here the work of scientists such as Martin Rees, James Lovelock, and Lynn Margulis, but also the Union of Concerned Scientists. The philosopher of science Jean-Pierre Dupuy addresses misplaced faith in science in *The Mark of the Sacred* (Stanford: Stanford University Press, 2013), and argues that a rediscovery of the sacred character of the world, and along with that an acknowledgment of the limits of human reasoning and the need for self-limitation, are essential to a viable future.

great cataclysmic event. It may take the form of an inexorable and mostly unnoticed "slow violence" that systematically undermines the health of all life.[14] Or it may be the remote-controlled, tele-murder violence that is "without hatred" that governs today's military operations.[15] However understood, it seems that we are in need of a new *theōria*, a new way of seeing the world that might better enable people to cherish the world and live more faithfully within it.

A Christian Way of Seeing?

Is there something like a uniquely Christian way of looking at the world? When followers of Christ look at the world, what do they see, and therefore also understand to be there? Put slightly differently, how does the *askēsis* or discipline of Christian living—living that is patterned after Christ's own way of being in the world—give rise to a *theōria* or hermeneutical consciousness that opens and focuses the world in new ways, enabling people to determine significance and meaning, fallenness and flourishing, in fresh ways?

One way to begin is to say that Christians see the world to be God's *creation*. It is the work of God's hands and the expression of God's love and delight (one can wonder if the parson Malthus had any inkling of this). But this can only be a beginning because what is needed is a rigorous development of what it means to say that we live in a created world, and then also a description of the manner by which this kind of seeing becomes possible. Recalling that *theōria* is always accompanied by an *ethos*, what ways of being in the world are prerequisite to seeing the world *as creation* rather than, perhaps, one of the many expressions of the world *as nature*?

It is important to ask this question because many Christians assume that

---

14. See the work of Rob Nixon in *Slow Violence and the Environmentalism of the Poor* (Cambridge, MA: Harvard University Press, 2011).

15. In *Hiroshima ist überall* (*Hiroshima Is Everywhere*) Günther Anders says, "The fantastic character of the situation quite simply takes one's breath away. At the very moment when the world becomes apocalyptic, and this owing to our own fault, it presents the image . . . of a paradise inhabited by murderers without malice and victims without hatred. Nowhere is there any trace of malice, there is only rubble. . . . No war in history will have been more devoid of hatred than the war by tele-murder that is to come. . . . [T]his absence of hatred will be the most inhuman absence of hatred that has ever existed; absence of hatred and absence of scruples will henceforth be one and the same" (quoted in Dupuy, *The Mark of the Sacred*, 194).

there is little difference between a world interpreted as creation and a world interpreted as nature. For some the world is what it is, with the key difference being that for Christians nature has its origin in God. In other words, the natural world becomes a created world the moment God is positioned at the beginning as the One who got it all going. God put in place the natural laws that keep the world functioning in the regular patterns that it does. Every once in a while, however, God is thought to intervene in a special way by interrupting, suspending, or perhaps even abrogating natural laws so as to produce a miracle.

This more or less deist characterization of creation is a profound mistake. Why? Because it does not at all reflect a biblical understanding of the world as the material place in which God's love is continually at work nurturing, healing, reconciling, and liberating creatures into the fullness of their being. As the psalmist puts it, God continuously and intimately faces the world, breathes upon it, because without God's animating Spirit the whole of life collapses into dust (Ps. 104:27–30). Focusing exclusively on origins ignores the fact that in scripture creation is as much about the salvation and the final consummation of things as it is about their beginning: protology, in other words, is inseparable from eschatology. That things come from God is important, but so too is the affirmation that creatures should move toward God because it is only in God that creatures find their fulfillment and their true end. More fundamentally, however, is the fact that a deist characterization of the world has no room for creation understood as the action of the triune God. Creation, rather than being a single event that happened a long time ago, signifies God's ongoing involvement in an economy and ecology that joins creaturely life with the life of God.[16] As such, the doctrine of creation is about the *character* of the world, about the way things now are and how they could be if they were fully participating in God's rule. Creation names a moral and spiritual topography of creatures called to be responsive to each other and to their Creator.[17]

16. Paul M. Blowers makes this point in a magisterial way in *Drama of the Divine Economy: Creator and Creation in Early Christian Theology and Piety* (Oxford: Oxford University Press, 2012). See also Denis Edwards's essay "Where on Earth Is God? Exploring an Ecological Theology of the Trinity in the Tradition of Athanasius," in *Christian Faith and the Earth: Current Paths and Emerging Horizons in Ecotheology*, ed. Ernst M. Conradie, Sigurd Bergmann, Celia Deane-Drummond, and Denis Edwards (London: Bloomsbury T&T Clark, 2014).

17. I developed this position in *The Paradise of God: Renewing Religion in an Ecological Age* (New York: Oxford University Press, 2003).

Equally important, this deist rendering ignores the fact that Christian theologians from early on advocated a Christian *theōria physike* or manner of seeing that enabled people to perceive the world as the place where God is intimately at work (in doing so they adapted and modified ancient philosophical forms of *theōria physike* that did not share the biblical understanding of the world as God's creation).[18] But to engage in this form of *theōria* it was essential that people practice the discipleship or *askēsis* that purifies seeing of the passions that distort the world and reduce it to the satisfaction of human desires. Put most directly, to see the world in a Christian manner is to see everything as God sees it. It was considered important for Christians to develop this way of seeing so that the world could be engaged faithfully and in a manner that brought healing to creatures and honor to God.

How is it possible for people to see the world this way, especially given the assumption that people are creatures and not the Creator? The answer: people can learn to see as God sees insofar as they become disciples of Jesus Christ and submit to the power of the Holy Spirit that enables them to participate in Jesus's *ethos*, his ways of being in the world.

Among early Christians it became a bedrock position that God bridged the chasm between Creator and creation in the incarnation of Jesus Christ. The eternal divine life and order took up residence in the person Jesus of Nazareth.[19] The Prologue to John's Gospel gave this memorable expression by describing Jesus as the divine, creating Word or *Logos*: "All things came into being through him, and without him not one thing came into being. What has come into being in him was life, and the life was the light of all people" (John 1:3-4). John's Gospel, however, was hardly unique in this regard. The early Christian hymn in Colossians spoke similarly of Christ: "He is the image of the invisible God, the firstborn of all creation; for in him all things in heaven and on earth were created, things visible and invisible, whether thrones or dominions or rulers or powers—all things have been created through him and for him. He himself is before all things, and in him all things hold together" (Col. 1:15-17). In the letter to the Hebrews Jesus

18. It was a general principle among ancient Greek philosophers that the task of thought was to bring the thinker into union with what is. A properly ordered soul is at its best when it is in harmonious alignment with the order of the world. Joshua Lollar describes Greek *theōria physike* in detail in Part I of *To See into the Life of Things: The Contemplation of Nature in Maximus the Confessor* (Turnhout, Belgium: Brepols, 2013).

19. Richard Bauckham has developed this theme in a detailed way in *Jesus and the God of Israel: God Crucified and Other Studies on the New Testament's Christology of Divine Identity* (Grand Rapids: Eerdmans, 2008).

is described as God's Son, the one who is "heir of all things" and the one "through whom he also created the worlds" (Heb. 1:2). And in the first letter to the Corinthians Paul describes Jesus Christ as the one Lord "through whom are all things and through whom we exist" (1 Cor. 8:6).

Passages like this make it abundantly clear that the earliest Christian communities understood creation in a decidedly Christological way. Jesus's body and life were understood to be the incarnation of the very life that God is, a life that creates and loves and nurtures and heals and reconciles all things that it touches. Jesus shows definitively that for God to create is also for God to redeem. New Testament scholar Sean McDonough summarizes it this way:

> The mighty works of Jesus, his proclamation of the kingdom of God, and the climactic events of the crucifixion and resurrection, clearly marked him as the definitive agent of God's redemptive purposes. But these mighty works could scarcely be divorced from God's creative acts. The memories of Jesus preserved in the gospels depict a man who brings order to the threatening chaotic waters, creates life out of death, and restores people to their proper place in God's world.[20]

Jesus is not simply a moral teacher. In his embodied life and way of being, in the various ministries he performs, he heals and restores creatures so that they can live the abundant life God has wanted them to live all along. Jesus is the interpretive key that allows us to unlock the meaning and significance of everything that is. His miracles, rather than being an interruption of the laws of nature, are acts of liberation that free people from the destructive bondages of demon-possession, hunger, illness, alienation, and death. Jesus is the complete, embodied realization of life's possibility as a way of love. To see him is to see the divine love that created the heavens and the earth. To participate in his life is to take on his point of view, and thus see everything in a completely new way. As Paul says, to be in Christ means that we no longer see others from a human point of view: "if anyone is in Christ, there is a new creation: everything old has passed away; see, everything has become new!" (2 Cor. 5:17). Put succinctly, Jesus is the hermeneutical lens that brings the world into the kind of focus that enables us to see it as either fallen or flourishing.

---

20. Sean M. McDonough, *Christ as Creator: Origins of a New Testament Doctrine* (Oxford: Oxford University Press, 2009), 2.

Theōria Physike in Maximus the Confessor

The originality and the wide-reaching implications of the biblical insight that in Jesus Christ a new way of seeing the world came into being took many years to develop. One particularly important place, however, was in monastic and mystical traditions that stressed and taught ascetical disciplines as a way to share in the divine life and the divine way of seeing all of reality. For the purposes of this chapter I will focus on the seventh-century Byzantine monk Maximus the Confessor because it is in him that we find Christian *theōria* developed in a rigorous and fruitful manner.

At the center of Maximus's thought is the conviction that in the incarnation of God in Jesus Christ the complete meaning of humanity and the world reaches its fulfillment, because in him we see the union of the divine and human nature. Maximus says that with Jesus "a wholly new way of being (*kainoterou tropou*) human appeared. God has made us like himself, and allowed us to participate in the very things that are most characteristic of his goodness."[21] Christ is the center of the universe and the gate through which true and complete life moves because in him we find the definitive expression of the eternal love that is life's beginning, sustenance, and end. If we Christians understand the world as fallen it is because of what Jesus enables us to see. Christ enables us to know when creation is achieving its godly end or purpose and when creatures have fallen away or fallen short of what God intends for them.

Maximus believed that Christians are called to play a mediating role in the created world, a role that helps fellow creatures move into the fullness of their life in God. They are, as Paul put it, to be "ministers of reconciliation" in the world (2 Cor. 5:18). To do this, however, requires that Christians learn to see the world rightly. They need the proper *theōria*. For Maximus, having the proper *theōria* means learning to see the divine *Logos* in the *logoi* of all created things. If Christ, as the scriptures attest, is the one through whom and in whom and for whom all things come to exist, if it is indeed the case that all things hold together through him, then he is the *Logos* that is present to each thing, informing its own *logos* or way of being.

Logos is a Greek term notoriously difficult to pin down because of its wide usage in ancient philosophical and spiritual contexts. As employed

---

21. St. Maximus the Confessor, "Ambiguum 7," in *On the Cosmic Mystery of Jesus Christ*, trans. Paul M. Blowers and Robert Louis Wilken (Crestwood, NY: St. Vladimir's Seminary Press, 2003), 70.

by Maximus it is fairly clear that it refers to something like the dynamic principle of order and coherence that enables a thing to be and become the unique thing that it is. Each thing, whether alive or not, is the realization of particular capacities. Insofar as a thing is prevented from achieving its potential is also the extent to which it can be said that its *logos* is being derailed, distorted, or denied. Creatures fail and fall because their *logos* is not in alignment with the divine *logos*.

At the heart of Maximus's creation theology we find an extended treatment of how God creates each creature with a unique *logos* enabling it to be the unique creature that it is. Christ is the eternal *Logos* continually and intimately present to each particular created *logos* as the power leading it into ever-greater communion, until finally complete communion is attained when God is all in all. No creature, however, is complete in itself. All creatures, we can say, are created to be in relationship with each other because they are the material expression of a triune love in perichoretic relationship. Creatures most fully become themselves by being in nurturing relationship with others. At the same time, the webs of creation are strengthened insofar as each creature is strong and best able to contribute to the health of the whole. When a creature's *logos* ceases to be in alignment with the divine *Logos*, that is, when a creature ceases to move harmoniously within the ways of God's communion-building love, is precisely the moment when it becomes possible to speak of creaturely fallenness.

Maximus offers a breathtaking vision in which not only humanity but the whole of the created world is invited to participate in the divine life of love, because it is only in God that created things can be properly known and seen for what they truly are: material expressions of love.[22] The incarnation of God in Jesus Christ means, "Man is made God by divinization and God is made man by hominization. For the Word of God and God wills always and in all things to accomplish the mystery of his embodiment."[23] To be formed by Christ means that our seeing of things cannot remain at a surface or superficial level. To see deeply is also to see the love of God that is at work in us, leading us into the fullness of life that is our unique possibility.

There is in Maximus the profound realization that each creature *in its*

---

22. Lollar gives beautiful expression to this vision when he says "God is moved by His love for creation and this motion is realized in the Dionysian outpouring of Goodness to beings and Its return, which is the very outpouring and return of God from Himself to Himself; hence the language of 'self-motion (*autokinesis*).' Everything that exists just is the 'motion' of God proceeding from Himself and returning to Himself" (*To See into the Life of Things*, 283–84).

23. "Ambiguum 7," 60.

very physicality is the intimate expression of God's love. All attempts that denigrate materiality as the realm to be abandoned or left behind would amount to a rejection of the incarnation of God in Jesus and a denial that Jesus is to be affirmed as fully human and fully divine. Lars Thunberg says, "[T]he presence of the Logos in the *logoi* is always seen as a kind of incarnation—a parallel to the incarnation in the historical Jesus—and thus an act of divine condescension."[24]

It is important to pause to notice the radical character of Maximus's vision. For a number of mystical theologians (especially those heavily influenced by Platonic traditions), as well as numerous theologians through the ages, the highest achievement of Christian life has required that embodiment and physicality must finally be left behind to achieve union with God. We might say that these theologians have lost their nerve before the radicality of the incarnation. Maximus rejects this approach because, to put it simply, Jesus did not have to abandon human embodiment to fully express the divine life. He did not have to shun creatureliness to be the Creator because he fully realized both within himself in a mysterious union that affirmed mutual indwelling and unconfused distinctness. The logical outcome of this position is that creation cannot ever be denigrated or despised. Being the material manifestation of the divine *logos* of love, creation is the home of God (cf. Revelation 21–22 where we are told that the everlasting home of God is among mortals).

God is intimately present to each creature as the source of its life, but not in such a way as to prevent creatures from being themselves. As Cornelia Tsakiridou states,

> A God who himself exists in a self-communicating manner, in Trinity, engages in conversation with his creatures, one by one and all together, and they in turn exist in order to converse with him their own existence, to be themselves and with each other, in his own life. He moves right inside their being to give it its very own mind, voice and life, to bring the finite beyond its finitude and into his life of eternity.[25]

Following this formulation, there is a dialogical relationship between God and creatures such that in being open to another and for another the fullness

---

24. Lars Thunberg, *Microcosm and Mediator: The Theological Anthropology of Maximus the Confessor*, 2nd ed. (Chicago: Open Court, 1995), 76.

25. Cornelia A. Tsakiridou, *Icons in Time, Persons in Eternity: Orthodox Theology and the Aesthetics of the Christian Image* (Burlington, VT: Ashgate, 2013), 176.

of life is approached. There is in this account an understanding of God's creative love (which is meant to find expression throughout the entire universe) as fundamentally an expression of hospitality: God creates the space and all the sources of nurture for creatures to come into his life and be strengthened to live the life they are uniquely prepared to enact.

Christian *Askēsis* on the Way to Seeing

So far we have been describing Maximus's vision of the world in which each creature gives expression to a *logos*, what we might also call the principle of intelligibility and order that makes it the unique thing that it is.[26] No *logos*, however, is self-subsisting or self-originating. It has its origin, sustenance, and end in the divine *Logos* who is Jesus Christ, which is to say that creaturely *logoi* exist only because of the will of God that desires and loves them into being. Creatures achieve the fullness of their being when they maximally participate in the divine *Logos* that is the meaning of the whole universe. In other words, the truth of each particular creature is realized when its *logos* is in harmonious alignment with the *Logos* that holds the universe together.

But some creatures, owing to their freedom, can be in more or less alignment with the *Logos*, which means that they can in refusing Christ also be out of alignment with their own *logos*. This point is important to underscore because when humans sin they contribute to what we might call the disalignment of various *logoi* in the world (as when human greed degrades an ecosystem such that creaturely life is frustrated in its ability to live into life's fullness). Using the language of Paul as expressed in Romans 8, we can say that human sin becomes the violating presence in the world that subjects creatures to various states of groaning and futility. Sin becomes a (cosmic) power in the world that makes it difficult for human and nonhuman creatures to realize their fulfillment in God, which is why sin must be addressed and corrected first so that creatures can freely move into their divinely appointed purpose. Sin and fallenness go together.

Maximus is clear that we can only understand things and ourselves in a Godly way when the movements of our entire life—the movement of our minds, the ordering of our affections, the practices of our bodies—are

---

26. Maximus writes, "all created things are defined, in their essence and in their way of developing, by their own *logoi* and by the *logoi* of the beings that provide their external content. Through these *logoi* they find their defining limits" ("Ambiguum 7," 57).

brought into conformity with Christ: a Christian *theōria* needs a Christian *askēsis*.[27] Christian discipleship is the key to the right ordering of ourselves and the right ordering of our vision so that we can see each other and everything as God sees it. In the incarnation of Jesus Christ God entered and "maintained the *logos* of creaturely origin while also wisely restoring humanity's means of existing to its true *logos*."[28]

God does not alter human nature by making it something else. Rather, God in Christ changes "the mode and domain of action proper to their nature."[29] God does not desire creatures to be something other than what they are. God only ever asks creatures to be themselves fully, a capacity that has become clouded and distorted because of sin. This means that as Christ leads people into the truth of their humanity he also at the same time leads them into a position to understand the truth of the world. Discipleship is what enables people to see each thing as the creature of God that it is, and to discern how and to what extent creatures are being prevented from realizing the fullness of the divine love that is at work within them.

With the sin of Adam, human freedom changed from good to evil. Evil is a misuse of freedom so that people fail to direct their energies in ways that are in alignment with God's will for things. Jesus, however, is the New Adam who reverses the movement from evil to good. The movement from evil to good, which is also a movement from corruption to incorruption, goes by the name of deification because it is the creature's appropriate, proportionate participation in the life of God. Our becoming God is not the result of our own effort. It is always only ever a gift of God's grace and God's loving invitation to lead people to their true end in him.

Maximus describes divinization or *theosis* as a process in which our spirit—the animating power of our life—is wholly given over to God's Spirit. "God becomes to the soul (and through the soul to the body) what the soul is to the body, as God alone knows, so that the soul receives changelessness and the body immortality; hence the whole man, as the object of divine action, is divinized by being made God by the grace of God who became man.

---

27. Maximus believes that in the paradise of the Garden of Eden something like a proper *theōria physike* obtained. With the "Fall" humans lost the ability to see each thing in terms of its reference and grounding in God. Adam's great mistake was to try to know the world by sensation alone rather than in terms of the divine love at work within it. We could say that the Fall represents a disordering or the making irrational (*alogos*) of a world that is meant to hold together because of the *Logos* of Christ.

28. "Ambiguum 42," in *On the Cosmic Mystery of Jesus Christ*, 82.

29. "Ambiguum 42," 90.

He remains wholly man in soul and body by nature, and becomes wholly God in body and soul by grace."[30] Without this process of growing into the likeness of God the virtues that enable people to see in a Godly way are impossible. With this realization the link between *theōria* and the proper *ethos* is established. To see the world as God's creation people must become creatures who live in Christ (recalling Paul's admonition to live by and exhibit the fruit of the Spirit [Gal. 5:22–25] and his succinct formulation that in baptism's crucifixion with Jesus "it is no longer I who live, but it is Christ who lives in me" [Gal. 2:20]). Living in Christ is the action that enables people to see the divine *Logos* at work in each creaturely *logos*. Seeing the divine *Logos* in things, in turn, makes it possible for Christians to participate in the healing of fallenness through the various ministries of reconciliation that Christ makes possible.

To live in Christ we must look to Christ to see what he does and what he accomplishes because it is in his action that we discern how he sees all that he meets. The gospels reveal Jesus to us as above all the one who is for others. Upon meeting another person Jesus sees first and foremost a child and a gift from God. What he most desires is that each creature be liberated to live the life that God has given it. His ministries of forgiveness, healing, exorcism, feeding, companionship, and reconciliation demonstrate that we live in a world where people are in bondage to forces of violence and hatred, illness and hunger, alienation and isolation. Jesus comes to free humanity from these forces so that all creatures can experience the love of God.[31] Jesus reveals that the goal of life is communion with each other and with God. In this communion life the relationships between creatures and God are fully healed so that each creature achieves what John called abundant life. Put in its most succinct formulation, we could say that Jesus reveals the truth of life as the movement of love.[32] His life from beginning to end, from crucifixion to resurrection, demonstrates the nature and the aims of divine love. This means that to achieve something like a Christian *theōria physike*, Christians must practice the Christian *askēsis* Jesus reveals in his own life. To love like Jesus is to perceive and engage the world the way that he did.

In an arresting passage that links the movement of Christ's own life with the life of the whole world, Maximus says,

---

30. "Ambiguum 7," 63.

31. In Romans 8, Paul argues that Christ's liberating work extends to the whole of creation so that every creature will know the love of God.

32. I have developed this theme in *Way of Love: Recovering the Heart of Christianity* (San Francisco: HarperOne, 2016).

The mystery of the Incarnation of the Word bears the power [*dynamin*] of all the hidden meanings and figures of Scripture as well as the knowledge of visible and intelligible creatures [*ktismatōn*]. The one who knows the mystery of the cross and the tomb knows the principles of these creatures. And the one who has been initiated into the ineffable power of the Resurrection knows the purpose [*skopon*] for which God originally made all things [*ta panta*].[33]

Here Maximus shows that a proper *theōria* requires of the believer an immersion into the history of God's economy as it is communicated through the scriptures and the world, God's two books. We cannot know the significance of what we see apart from the sacred drama that is revealed through the Word of God, nor can we know the purpose of things apart from Christ. The eternal *Logos* acts like the interpretive lens that allows us to see the *logoi* of created things as signifying God's blessing and God's love.

Maximus then adds that for us to see this way, our hearts and minds must go through a crucifixion experience because it is there that the purification of the ego occurs so that we can see things in the light of God's love rather than the distorting, dissimulating clouds of our own self-serving passions.

All visible things [*phainomena*] need a cross, that is, a capacity that restrains the affection for them on the part of those who are sensibly attracted to them. And all intelligible things demand a tomb, that is, the complete immobility of those who are intellectually inclined toward them. For when natural activity and movement are removed along with the inclination for all these things, the Logos, who is alone self-existent, reappears as though he were rising from the dead, circumscribing everything that originates from him.[34]

In a manner reminiscent of Paul's description of baptism as the believer's old self being crucified with Christ so that he or she might also be resurrected into newness of life (Rom. 6:3–14), Maximus is describing a process in which our vision and our understanding are cleansed and our priorities reoriented so that the life we live is now in conformity with the life God has intended all along. Living this cruciform life, a life in which love directs us to

---

33. Maximus as quoted in Tsakiridou, *Icons in Time, Persons in Eternity*, 179.

34. Here following the translation of Maximus by Blowers in *Drama of the Divine Economy*, 362.

seek the good of others rather than pleasures for ourselves, we come to see everything in God. We come to see that each thing is the unique expression of God's love and exists for no other reason than to give glory to God as the giver and nurturer of its life. The essential task is to learn to love properly, for it is in the mode of divine love that the human presence on earth becomes one that heals and reconciles all things in their individual being and in their life together.

One could say that learning to love properly is the heart of the Christian task because, as Maximus and numerous spiritual writers have insisted, improper self-love, what ascetic writers call life according to the passions, so easily gets in the way. Improper self-love is reflected in a lustful or pornographic relationship with things, a relationship in which things signify or matter primarily in terms of what they can do for us. When in a lustful relationship with another the integrity of that other, and thus also the course of life that would fulfill it, is denied because its life is now made to serve my own.[35] This is why a life lived according to the passions—traditionally seven in number: gluttony, unchastity, avarice, anger, dejection, listlessness, and pride—leads to the tyranny of creatures and the degradation of the whole created world.[36] Life lived according to the passions renders Christian *theōria* impossible. It is only love for others that enables people to see the world and its creatures as God does.

Maximus describes the passions as an irrational attachment to the body. It is important to stress that Maximus does not reject or despise bodies in and of themselves. This he could not do since each body is the material manifestation of God's love. "It is not the body itself, nor the senses nor the possible faculties themselves which are evil, but only their wrong use. . . . Self-love is defined as love for the body, not because the body is linked with evil, but because attachment to the body prevents man's entire attachment to his divine end."[37]

35. Tsakiridou gives the following helpful summary: "When creatures are perceived spiritually or in a God-loving manner (*theophilos*), they are seen in their true nature and subsistence, as his living (incarnating) works. When, by contrast, they are perceived from the standpoint of desire or self-love (*philautia*), this vital, animating reality in them disappears and the mind imposes its own self-serving and distorted reasons . . . on things. . . . The passions obscure the inherent divinity and sanctity of creation and it is therefore in their activities rather than in the things themselves that evil arises" (*Icons in Time, Persons in Eternity*, 183).

36. Dumitru Staniloae, one of the last century's leading interpreters of Maximus, gives a useful account of the passions in *Orthodox Spirituality: A Practical Guide for the Faithful and a Definitive Manual for the Scholar* (South Canaan, PA: St. Tikhon's Seminary Press, 2003).

37. Thunberg, *Microcosm and Mediator*, 247–48.

When our focus and attention rests on the material body alone we forget both the *logos* that is interior to that thing directing it to its fulfillment, and we forget the divine *Logos* in which it participates and that is leading it to its eternal well-being in God, because what has become most important is how that thing can be made to serve our end. A passionate embrace of the world, we could say, is invariably a superficial and a destructive looking at others because it does not see God's love of creatures everywhere at work.[38] Things are degraded and destroyed because their movement, rather than contributing to the flowering of God's whole creation, has been channeled to suit the narrow aim of human ambition.

This account shows that asceticism, the *askēsis* that informs the *ethos* that makes possible a Christian *theōria physike*, has nothing to do with the denial or denigration of the material world. Genuine asceticism leads to the purification and intensification of Christ-like love that leads to the world's healing and reconciliation. Without this love the world remains wounded and fallen.

The passions are irrational (*alogos*), which means that they work contrary to the divine *Logos* that is constantly present to each creature leading it into the fullness of its own life and its life together with everything else. As disciples of Jesus Christ and as members of his body, Christians have the high calling to become agents of the Holy Spirit's work of healing and celebration. When human hearts are inspired by Jesus Christ, then his divine *Logos* takes hold of our own *logos* so that it shares in the mediating work that is the work of the incarnation: "Things that are by nature separated from one another return to a unity as they converge together in one human being. When this happens God will be *all in all* (1 Cor. 15:28), permeating all things and at the same time giving independent existence to all things in himself. Then no existing thing will wander aimlessly or be deprived of God's presence."[39]

Christ is the hermeneutical key that enables Christians to understand

---

38. Maximus does allow for "good passion" insofar as it has been made captive in obedience to Christ ("Ad Thalassium 1," in *On the Cosmic Mystery of Jesus Christ*, 98).

39. "Ambiguum 7," 66. Just as in Christ the unity and difference of natures is maintained, so too in creation. Each member of creation is distinct but is now brought into a mutuality of relationship that strengthens each one and the whole. Christ reveals that it is not difference but division that is the problem besetting our world. The work of humanity, inspired and patterned as it is on Christ, is to honor difference but reconcile division. See the discussion of Thunberg in *Microcosm and Mediator*, 65, and Maximus's account of how Jesus mediates and heals division in "Ambiguum 41" in Andrew Louth's *Maximus the Confessor* (London: Routledge, 1996), 155–62.

that creatures are fallen because they fail to participate fully in the divine love that creates, sustains, and celebrates the whole world. Christ is also the inspiration that guides Christians into the various ministries of reconciliation that lead creatures out of their fallen condition so that they can live into the telos or goal that God has desired for them from the beginning.

PART IV

Reimagining the Conversation:
Faithful Ways Forward

9 The Fall of the Fall in Early Modern Political Theory

The Politics of Science

WILLIAM T. CAVANAUGH

There is a general cultural assumption in the West that any antagonism between science and theology is inherent in scientific method. The great secularizer is science, it is often thought, because the fancy or sheer unprovability of theological beliefs eventually runs into scientific fact. We can no longer take the Fall story seriously, for example, because it is just a story. So the story goes.

For those who are invested in a more fruitful dialogue between science and theology, however, it is helpful to know that secularization is not the inevitable result of science. And one way of loosening the grip of the story told above is to show that secularization—including the secularization of science—has causes that are nonscientific. Max Weber, for example, thought that capitalism, not science, was the great secularizer.[1] Charles Taylor, Brad Gregory, and others have shown that secularization has theological roots.[2]

In this chapter I offer a contribution to such efforts by tracing a political genealogy of secularization through the fate of the Fall in early modern

---

1. Max Weber, *The Protestant Ethic and the Spirit of Capitalism*, trans. Talcott Parsons (London: Routledge, 2001).

2. Charles Taylor, *A Secular Age* (Cambridge, MA: Harvard University Press, 2007); Taylor argues that secularization is a contingent byproduct of certain reform movements within late medieval and early modern Christianity that relocated the holy from the external world to the interiority of a new kind of human self. Brad S. Gregory, *The Unintended Reformation: How a Religious Revolution Secularized Society* (Cambridge, MA: Harvard University Press, 2012); Gregory argues that the splintering of Christian unity in the Reformation ultimately caused a turning away from public Christianity in the West.

political theory. What I hope to show is that the eclipse of the Fall has roots that are political, not scientific, and that the "naturalization" of the Fall in early modern political theory contributes to the rise of the modern state and to the divorce between theology and political science and between theology and natural science. I begin by giving a brief overview of the importance of the Fall in premodern Christian political thought. I then examine the fate of the Fall in early modern thought, briefly discussing Niccolò Machiavelli and Francisco de Vitoria, but concentrating on the English tradition most influential in our context, namely Thomas Hobbes, Robert Filmer, and John Locke. I show how and why the Fall is replaced by the "state of nature" as prehistorical justification of political power. I conclude with some comments on the genealogy of the relationships among science, politics, and theology. This chapter considers what is lost in all three areas when Western society no longer uses the Fall to mark the difference between the way things are and the way things are meant to be.

The Fall in Medieval Political Theory

The biblical narrative of the Fall occupied a foundational place in traditional Christian political theory. The Fall was seen as either the reason that coercive government was necessary or a significant factor affecting what was possible in human government. This emphasis on the Fall should not be misunderstood, however, as an example of Christian pessimism about human nature that has been overcome in more secular societies. In traditional Christian political theory, the primary referent of human nature is a pre-Fall phenomenon; human nature consists of the capacities instilled by God in humans at creation. The consensus among patristic and medieval commentators was that human beings are by nature sociable creatures that are inclined to love their fellows. Despite the opinion commonly voiced a half-century ago that Christianity owed its view of humanity's natural sociability to the reintroduction of Aristotle to the West in the thirteenth century, scholarly consensus now recognizes that Christian thinkers from Lactantius onward acknowledged the natural sociability of humankind.[3] Augustine, for exam-

3. Cary J. Nederman, "Nature, Sin and the Origins of Society: The Ciceronian Tradition in Medieval Political Thought," *Journal of the History of Ideas* 49, no. 1 (January–March 1988): 3. For an earlier recognition of the patristic and medieval consensus on human sociability, see A. J. Carlyle, *A History of Mediaeval Political Theory in the West* (Edinburgh, London: William Blackwood, 1950), 1:125ff. See also Gaines Post, *Studies in Medieval Legal*

ple, writes that "since every person is a part of the human race, and human nature is social, each person also has a great and natural good, the power of friendship."[4]

As Augustine also says, however, "The human race is, more than any other species, at once social by nature and quarrelsome by perversion."[5] The distinction between nature and the perversion of that nature is crucial for Augustine and for the Christian tradition as a whole. Augustine talks about God's intention in Adam to bring forth a multitude from this one individual and thereby "teach mankind to preserve a harmonious unity in plurality." At the same time, Augustine says, in Adam God laid the foundation of the two cities, those who would join the evil angels in their punishment, and those who would share the company of the good angels in their reward.[6] In the story of Adam, God teaches us both what we ought to be and what we have come to be because of human choice. The story of the Fall, therefore, is not simply a claim about the evil that lurks in human souls and the necessity of coercive government to make human social life possible. The Fall is also a lesson about the way humans ought to be and behave, the *telos* of human life, based on the way that humans *really* are, the way that they were created by God. The Fall is not simply a pessimistic doctrine, but, on the contrary, gives humans hope that the evil that people do to one another is not natural, that is, is not inscribed in the way things are from creation, and is therefore not simply inevitable.

Though Christian thinkers saw human coercive government as instituted by God, the patristic to high-medieval consensus was that coercive government is not natural, but rather a divine response to human sin.[7] Ire-

---

Thought: Public Law and the State, 1100–1322 (Princeton: Princeton University Press, 1964), 494–561.

4. This is my translation of "Quoniam unusquisque homo humani generis pars est, et sociale quiddam est humana natura, magnumque habet et naturale bonum, vim quoque amicitiae," Augustine, *De Bono Conjugali* 1, quoted in Carlyle, *A History of Mediaeval Political Theory in the West*, 1:125n3.

5. Augustine, *City of God* 12.28, trans. Henry Bettenson (Harmondsworth, UK: Penguin, 1972), 508.

6. Augustine, *City of God* 12.28 (p. 508). Augustine says that the foundation is laid in God's foreknowledge; the origin of the two cities in human history for Augustine is in the story of Cain and Abel; see book 15 of the *City of God*.

7. "[T]he normal view of the Fathers is clear, namely, that while coercive government is not a 'natural' institution, and is a consequence of the Fall and related to men's sinful ambitions, yet it is also a divine remedy for the confusion caused by sin, and is therefore a divine institution"; Carlyle, *A History of Mediaeval Political Theory in the West*, 2:144. Some

naeus traces the subjection of humans to other humans to the Fall.[8] In 14.15 of the *City of God*, Augustine follows this line, and describes the "order of nature" in this way: God "did not wish the rational being, made in his own image, to have dominion over any but irrational creatures, not man over man, but man over the beasts."[9] The subjection of one human person to another came about, Augustine says, because of sin. "And yet by nature, in the condition in which God created man, no man is the slave either of man or of sin."[10] In this instance it is clear that "nature" refers to the condition of human beings before the Fall, not after; humans are created in a condition of natural freedom. Sin, and the need for coercive government to mitigate its effects, is not simply a given aspect of human life. This, indeed, is what distinguishes Genesis from the Babylonian creation myth, the *Enuma Elish*.[11] Both the Israelites and the Babylonians looked out upon the same world marked by violence and human evil. In the *Enuma Elish*, however, there is no Fall; things are messed up from the start. Evil is just part of the way things are. For the authors of Genesis, on the other hand, the Fall makes clear that there is nothing natural or foundational about human evil. The Fall, furthermore, establishes the possibility of viewing history eschatologically; if the way things appear to be is not the way things are meant to be, then there is hope that things may be radically changed. Augustine applies this eschatological view not only to sin and evil but to coercive government, which is only made necessary by sin. Augustine closes 14.15 by looking forward to the day when "all injustice disappears and all human lordship and power is annihilated, and God is all in all."[12] Augustine may not have expected that

commentators trace the emphasis in Christian political thought on the difference between pre-Fall innocence and post-Fall sin to the Stoic conception of a now-corrupted Golden Age that serves as the ideal basis of law; see R. A. Markus, "The Latin Fathers," in *The Cambridge History of Medieval Political Thought c. 350–c. 1450*, ed. J. H. Burns (Cambridge: Cambridge University Press, 1988), 98, and George Klosko, *History of Political Theory: An Introduction*, vol. 1 (Fort Worth, TX: Harcourt Brace, 1994), 152–56, 211.

8. Irenaeus, *Against Heresies* 5.24, in *From Irenaeus to Grotius: A Sourcebook in Christian Political Thought*, ed. Oliver O'Donovan and Joan Lockwood O'Donovan (Grand Rapids: Eerdmans, 1999), 16–18.

9. Augustine, *City of God* 14.15 (p. 874).

10. Augustine, *City of God* 14.15 (p. 875).

11. Alexander Heidel, ed., *The Babylonian Genesis: The Story of Creation* (Chicago: University of Chicago Press, 1963).

12. Augustine, *City of God* 14.15 (p. 875). Granted, this passage is ambiguous, because Augustine justifies God's providential institution of slavery and, following Paul, admonishes slaves to obey their masters until God brings this era to a close.

day anytime soon, but the eschatological view has the effect *in the present* of destabilizing and relativizing any human claim to political power.

Augustine's position was echoed in the medieval period prior to the recovery of Aristotle: Ambrose, Gregory the Great, Isidore of Seville, and Gregory VII (to name a few) all thought that coercive government was divinely instituted but not natural.[13] Human coercive power, in other words, was utterly dependent on the will of God and not embedded in claims about human nature. It was a remedy for sin made necessary because the Fall had damaged the natural sociability of human beings. Thomas Aquinas, however, pressed the origins of government behind the Fall of humankind. Citing Augustine's opinion from the *City of God* 19.15, Aquinas says that the mastery of humans over one another can be understood in two ways: as slavery and as the direction of one "towards his proper welfare, or to the common good."[14] The former could not have existed in the pre-Fall state of innocence, according to Aquinas, because it implies the surrender of freedom and the infliction of pain, but the latter is proper to the state of innocence. Political community is *natural* in the sense that it corresponds to the intended—that is, created—end of human life, which is life together with others and with God.

Although Aquinas locates the origin of government in the state of innocence, the Fall nevertheless continues to exert an important influence in Aquinas's theory of politics. In the fallen state, government is needed to "make men good,"[15] and this requires coercive government.[16] Although Aquinas departs from Augustine in assigning government to the state of innocence, there is an important continuity between the two. As Janet Coleman writes, pre-Fall government in Aquinas is directive, not coercive.[17] Al-

---

13. Ambrose, "Letter to Simplicianus" (Letter 37), in St. Ambrose, *Letters*, Fathers of the Church 26, trans. Sr. Mary Melchior Beyenka (New York: Fathers of the Church, 1954), 286–303; St. Gregory the Great, *Pastoral Care* 2.6, trans. Henry Davis (Westminster, MD: Newman, 1955), 59–60, and Pope Gregory I, *Morals on the Book of Job* 21.23–24, trans. James Bliss (Oxford: J. H. Parker, 1845), 534–36; Isidore of Seville, *Sentences* 3.47, in O'Donovan and O'Donovan, *From Irenaeus to Grotius*, 206; Pope Gregory VII, "Letter to Hermann of Metz," 8.21, in *The Correspondence of Pope Gregory VII: Selected Letters from the Registrum*, trans. Ephraim Emerton (New York: W. W. Norton, 1969), 166–75.

14. Thomas Aquinas, *Summa Theologica*, I.96.4.

15. Aquinas, *ST*, I–II.92.1.

16. Aquinas, *ST*, I–II.90.3 ad 2.

17. Janet Coleman, *A History of Political Thought: From the Middle Ages to the Renaissance* (Oxford: Blackwell, 2000), 109. It should be noted that some medieval writers after Aquinas continued to see government as a result of the Fall. Marsilius of Padua, for exam-

though Aquinas is more sanguine about the capacities of human nature after the Fall, for both Augustine and Aquinas government only becomes coercive because of the Fall.

The Fall in Christian thought marks the divide between two kinds of nature. As Aquinas puts it, "Man's nature may be looked at in two ways: first, in its integrity, as it was in our first parent before sin; secondly, as it is corrupted in us after the sin of our first parent."[18] Considering our original, pre-Fall nature is clearly more than a historical exercise; it shows us what God's intentions for human life are, and it therefore marks the current disabilities of human nature as not simply inevitable or unfixable. Consideration of the pre-Fall state distinguishes between the way things are and the way things are meant to be.

The Rise of Leviathan and the Fall of the Fall

The eclipse of the idea of a Fall of humankind did not have to await the rise of evolution and the prestige of the natural sciences; it was eclipsed in the early modern attempts to create a new naturalistic science of politics. Niccolò Machiavelli is often considered the first modern European thinker to attempt to establish politics on some basis other than theology. Machiavelli was not only contemptuous of the influence of Christianity on politics; more broadly he sought to establish politics on an empirical basis, that is, on what *is* rather than on what should be. The Fall is simply absent from Machiavelli's political theory.[19]

The naturalization of politics in the early modern period was promoted not only by skeptics like Machiavelli, but by the Dominican priest Francisco de Vitoria. Vitoria established political authority not on God's providential grace but on God's law, expressed in the law of nature established in the act of creation. Vitoria followed the Aristotelian-Thomist idea that political societies were established on the basis of humanity's natural sociability. As Vitoria writes in his treatise *On Civil Power*, "the primitive origin of human

---

ple, writes, "Now if Adam had remained in this status [of innocence], the establishment or differentiation of civil offices would not have been necessary for him or for his posterity"; Marsilius of Padua, *Defensor Pacis* 1.6.1, trans. Alan Gewirth (Toronto: University of Toronto Press, 1980), 21.

18. Aquinas, *ST*, I–II.109.2.

19. One searches *The Prince* in vain for any mention of the Fall; Niccolò Machiavelli, *The Prince*, trans. William J. Connell (Boston, New York: Bedford/St. Martin's, 2005).

cities and commonwealths" is "a device implanted by Nature in man for his own safety and survival."[20] Vitoria's main target here is those who claim that the power of monarchs derives from human origin—the commonwealth or the people—but he also rejects the Augustinian idea that the state of innocence was one of freedom from the rule of other "men," with the implication that human government was only a later imposition due to sin.[21] Vitoria contends that "[e]ven if there were no scriptural authorities on this matter, reason alone would be able to resolve the question."[22] Vitoria considers and rejects the idea that the power of kingship changes after the advent of Christ; he does not even consider the possibility that the Fall makes any difference. The Fall is entirely absent from this discussion and indeed from his entire treatise *On Civil Power*. When Vitoria briefly mentions the Fall in his treatise *On the American Indians*, it is to refute Wyclif's and Fitzralph's idea that civil dominion was given to Adam and Eve and then lost because of their sin. For Vitoria, civil dominion does not depend on grace but is embedded in nature, and the Fall has no effect whatsoever on the legitimacy of kings.[23]

Rather than study the skeptical and Thomist traditions in the early modern period, where the Fall is essentially absent, I will concentrate my analysis on the English tradition—Hobbes, Filmer, and Locke—both because the English tradition has had the most profound influence on liberalism in the United States and Europe, and because Adam continued to be a significant, if altered, presence in the work of these three thinkers, and so they present the "hardest cases" for my thesis.

Thomas Hobbes is sometimes studied as the first "modern" political theorist, one who attempted to move the study of politics away from a basis in scripture and onto a more naturalistic basis. For Mark Lilla, for example, Hobbes's consideration of the anthropological origins of "religion" is a decisive turn away from theology, changing the subject from God to religion as a human phenomenon. Lilla credits Hobbes with launching the modern separation between religion and politics.[24] Likewise Ross Harrison con-

20. Francisco de Vitoria, *On Civil Power* 1.2, in *Vitoria: Political Writings*, ed. Anthony Padgen and Jeremy Lawrance (Cambridge: Cambridge University Press, 1991), 9.

21. Vitoria, *On Civil Power* 1.5 (p. 13). Vitoria rejects the Augustinian position as "madness" without naming Augustine.

22. Vitoria, *On Civil Power* 1.5 (p. 14).

23. Francisco de Vitoria, *On the American Indians* 1.2, in *Vitoria: Political Writings*, 241–43.

24. Mark Lilla, *The Stillborn God: Religion, Politics, and the Modern West* (New York: Knopf, 2007), 84–88.

tends that, although Hobbes makes extensive use of scripture in his work, its load-bearing role has shifted; in Hobbes's thought "the premises, argument, and conclusions would all still stand even if we were to remove God from the thought."[25] Hobbes attempted to make political theory into a proper science; he worked briefly under Bacon and is, as Harrison says, "best thought of as a participant in the modern scientific revolution."[26] A growing body of scholarship, however, contends that it is no longer plausible to ignore the latter half of Hobbes's *Leviathan* and its detailed scriptural exegesis or to regard them as Hobbes's way of placating the religious sensibilities of the age. Howard Warrender, A. E. Taylor, A. P. Martinich, and others regard Hobbes as a sincere, if idiosyncratic, Christian believer. Martinich has argued that Hobbes's main aims were to show that Christianity and science are compatible, and to show that Christianity could not legitimately be used to destabilize civil government.[27] Matthew Rose has argued that the politics of *Leviathan* are the politics of the Bible, which is, in Hobbes's view, all about the establishment of God's government on earth.[28]

Though I find the second type of interpretation convincing, I have no intention of trying to resolve the issue here. What is crucial for my present purposes is simply to show how Hobbes largely replaced the biblical story of the Fall with the notion of a "state of nature" as part of a narrative that justified the existence of political authority. Hobbes's attempt to establish a science of politics replaced the dual pre-Fall/post-Fall schema of human nature with a unitary account of nature. While use of the "state of nature" has certain affinities with medieval attempts to justify human government on the basis of human nature, there are significant differences, even where some account of the Fall continues to appear.

Hobbes's science of politics depends on an account of nature, most famously in his construction of a state of nature which, being a state of war, necessitates the artifice of the state to make life bearable. Whether or not the state of nature ever actually existed historically does not seem to interest Hobbes;[29] the state of nature is more a thought experiment in what

---

25. Ross Harrison, *Hobbes, Locke, and Confusion's Masterpiece: An Examination of Seventeenth-Century Political Philosophy* (Cambridge: Cambridge University Press, 2003), 54.

26. Harrison, *Hobbes, Locke, and Confusion's Masterpiece*, 58.

27. A. P. Martinich, *The Two Gods of* Leviathan: *Thomas Hobbes on Religion and Politics* (Cambridge: Cambridge University Press, 1992), 5.

28. Matthew Rose, "Hobbes as Political Theologian," *Political Theology* 14, no. 1 (February 2013): 23.

29. "It may peradventure be thought, there was never such a time, nor condition of

life would be like without coercive government. It depends therefore, on a conception of what human beings are naturally like when the artificial constructs of human civilization are stripped away. The picture is not pretty. The state of nature is not only a state of war, but one in which preemptive war is reasonable, given the need to protect one's life and property from the depredations of others. Reason, then, applies in the state of nature, and the natural law, being natural and not artificial, is meant to apply in the state of nature. The natural law is the law of God; the laws of nature are only genuine laws because God commands them.[30] But because the fear of other people is more immediate than the fear of God, the natural law applies only *in foro interno* until the coercive state can provide sanctions *in foro externo*.[31] As Hobbes says in *Leviathan*, "covenants without the sword are but words and of no strength to secure a man at all."[32] In the state of nature, then, the law of nature instills in each human being the desire to seek peace, as a means of self-preservation, but the law, being devoid of sanctions, awaits the creation of the coercive political authority to enforce it. We know by the law of nature that God commands that people keep covenants, but we know that it is dangerous to keep covenants unless others do so as well. It is therefore in our interests to create a state to threaten us all, *in foro externo*, into keeping our covenants.[33] Just because we leave the state of nature to form a government, however, we do not thereby become immune to the self-interested motives that make government necessary. Human nature does not change in the move from state of nature to the creation of civil government.[34]

Like the state of innocence in medieval political thought, the state of nature serves for Hobbes as a prehistoric condition that justifies civil gov-

---

war as this; and I believe it was never generally so, over all the world"; Thomas Hobbes, *Leviathan: Or the Matter, Forme, and Power of a Commonwealth, Ecclesiasticall and Civil* (New York: Macmillan, 1962), 101. Hobbes's "state of nature" is less a historical claim and more a hypothetical condition like John Rawls's "original position."

30. The opening line of Hobbes's *Leviathan* declares that nature is "the art whereby God hath made and governs the world" (19). Scripture is another form of the law of nature, which, when interpreted rightly by the use of reason, teaches what is contained in the natural law. So Hobbes writes of the scriptures, "As far as they differ not from the laws of nature, there is no doubt but they are the law of God, and carry their authority with them, legible to all men that have the use of natural reason: but this is no other authority than that of all other moral doctrine consonant to reason" (284).

31. Hobbes, *Leviathan*, 122–23. See also Martinich, *The Two Gods of* Leviathan, 136–37.

32. Hobbes, *Leviathan*, 129.

33. On this point, see Martinich, *The Two Gods of* Leviathan, 71–74.

34. Glen Newey, *Hobbes and* Leviathan (London: Routledge, 2008), 81–82.

ernment, though Hobbes's state of nature is the polar opposite of the state of innocence. The point, however, is not necessarily that Hobbes is more pessimistic about the human condition as it stands. The main difference is that for Hobbes the Fall plays no role. The state of nature describes neither a prelapsarian state of innocence (obviously) nor a postlapsarian state of fallenness from a formerly pristine state. As in the *Enuma Elish*, there is no looking back to an original goodness which thereby serves a normative function. The human being that Hobbes describes in *Leviathan* is not estranged from his or her true nature, from what he or she ought to be. The state of nature that Hobbes describes as a state of war is what Genesis describes as the consequences of human sin; humans become aware that they are not what they are meant to be, and therefore of their need for redemption. The Christian *foro interno* is the conscience that speaks the law written on human hearts, which points to the difference between what we are and what we should be (Rom. 7:23). The Christian conscience tells people they are unrighteous and need redemption; Hobbes's law of nature that speaks *in foro interno* tells people that they are unsafe and need coercive government to protect their self-interests. Hobbes's human subjects are dissatisfied with the state of nature, but not with themselves.[35]

Hobbes does express concern for the redemption of people from their sins. When Hobbes briefly discusses the sin of Adam in *Leviathan*, the effect of that sin on posterity is limited to mortality, the loss of eternal life, which Christ reverses.[36] But the redemption from the sin of Adam that Jesus works only has an indirect effect on Hobbes's politics. Hobbes's "artificial man" Leviathan owes a great deal to the theological concept of Adam and Christ as representative persons; just as all sinned through Adam, so all are offered salvation in Christ. The sovereign is the "artificial soul" of the "mortal god" Leviathan, a representative person established by covenant.[37] But Hobbes writes, "To make covenant with God, is impossible, but by mediation of such as God speaketh to."[38] As Christopher Hill comments, "Hobbes' object here is to substitute Leviathan for Jesus Christ. None represents God's person

---

35. On this point, see Paul D. Cooke, *Hobbes and Christianity: Reassessing the Bible in Leviathan* (Lanham, MD: Rowman & Littlefield, 1996), 105–9.

36. "Now Jesus Christ hath satisfied for the sins of all that believe in him; and therefore recovered to all believers, that eternal life which was lost by the sin of Adam" (Hobbes, *Leviathan*, 326).

37. Hobbes, *Leviathan*, 19–20. On the influence of Reformed covenant theology on Hobbes, see Martinich, *The Two Gods of* Leviathan, 143–50.

38. Hobbes, *Leviathan*, 109.

'but God's lieutenant, who hath sovereignty under God.'"[39] The effects of Adam and Christ on political history seem negligible. Hobbes divides history into four epochs. In the first, beginning with Adam, God rules over all humanity rather than a particular nation. With Abraham and then Moses, God enters into a sovereign-making covenant with the Chosen People. From Saul onward, the third period, human monarchies replace theocracy; there is no kingdom of God on earth, though the Christian sovereign is mandated to rule according to God's law implanted in nature and scripture. Only with the second coming of Jesus will the fourth period of history begin and the kingdom of God return. Jesus's kingdom when he walked the earth was not of this world; Jesus was not a king, but came to teach souls how to attain eternal life.[40] While we await the second coming, the Christian is to prepare for eternal life and obey his or her earthly sovereign. Neither the Fall of Adam nor the undoing of the Fall by Christ began a new epoch in world history; neither the first Adam nor the second Adam had any significant effect on the government of the world. One of Hobbes's primary goals is to ensure that Christianity could not be used to support sedition against civil government. One of the ways he does so is by ensuring that there is no tension in the present between God's rule and human rule, between the way things are and the way things ought to be.

The Fall is not entirely absent from Hobbes's work, but it serves a very different purpose for Hobbes than it did for medieval thinkers. In *De Cive*, Hobbes gives a brief account of the Fall narrative in Genesis, encapsulated in three verses: 2:15, "the most ancient of all Gods commands," against eating of the tree of the knowledge of good and evil; 3:5, *"Yee shall be as Gods, knowing good and evill"*; and 3:11, *"Who told thee that thou wert naked?"* For Hobbes, the significance of 3:11 is the following: "As if he had said, how comest thou to judge that nakedness, wherein it seemed good to me to create thee, to be shamefull, except thou have arrogated to thy selfe the *knowledge* of good and evill?"[41] The whole point of the episode for Hobbes is that Adam's sin consists in arrogating to himself the power to judge good and evil, which belongs solely to the king. "Before there was any government, *just* and *unjust* had no being, their nature onely being relative to some command, and every

39. Christopher Hill, "Covenant Theology and the Concept of a 'Public Person,'" in *The Collected Essays of Christopher Hill*, vol. 3 (Amherst: University of Massachusetts Press, 1986), 317. The internal quote is from *Leviathan*, chap. 18.

40. Hobbes, *Leviathan*, 325–27, 342–58.

41. Thomas Hobbes, *De Cive* 12.1, in *De Cive: The English Version* (Oxford: Clarendon, 1983), 147.

action in its own nature is indifferent; that it becomes *just* or *unjust*, proceeds from the right of the Magistrate."[42] Here Hobbes includes no discussion of the *foro interno*, which, in any case for Hobbes, is not the same as the ability to judge right from wrong; "private men" are guilty of assuming the role of the king when they try themselves to judge right from wrong. Indeed, I don't sin if I do what the king commands me to do, even if I think it is an unjust command. Anyone who makes his or her own judgment, whether obeying the king or not, sins.[43] The private knowledge of good and evil "cannot be granted without the ruine of all Governments."[44]

For Hobbes, coercive government is not made necessary by the Fall; it exists regardless of the Fall. The king stands in the same place that God stood in Genesis 2:15, commanding against private judgment. Later in *De Cive*, Hobbes alludes to a prelapsarian state in one passage: "In the beginning of the world God reigned indeed, not onely naturally, but also *by way of Covenant*, over *Adam*, and *Eve*; so as it seems he would have no obedience yeelded to him, beside that which naturall Reason should dictate, but *by the way of Covenant*, that is to say, by the consent of men themselves." Hobbes immediately adds, however,

> Now because this *Covenant* was presently made void, nor ever after renewed, the originall of Gods *Kingdom* (which we treat of in this place) is not to be taken thence. Yet this is to be noted by the way, that by that precept of not eating of the tree of *the knowledge of good and evill . . .* God did require a most simple obedience to his commands, without dispute whether that were *good*, or *evill*, which was commanded.[45]

Here Hobbes seems to be echoing the Reformed covenant theology of his day, which posited an original "covenant of works" with Adam by which humans would enjoy eternal happiness if they obeyed God's commands. Because of Adam's sin, the original covenant was made void, and a subsequent "covenant of grace" was made with Abraham and his descendants.[46] The important point for our purposes is that "the originall of Gods *Kingdom*" is not to be taken from the prelapsarian condition, which ceases to have relevance. The prelapsarian state does not stand as a model of what human life

---

42. Hobbes, *De Cive* 12.1 (p. 146).
43. Hobbes, *De Cive* 12.2 (p. 147).
44. Hobbes, *De Cive* 12.6 (p. 150).
45. Hobbes, *De Cive* 16.2 (p. 201).
46. Martinich, *The Two Gods of* Leviathan, 147–50.

is meant to be; the Fall story is rather a simple morality tale of disobedience and the arrogation of private judgment that, by nature, belongs to the king.

Filmer and Locke

Sir Robert Filmer was rescued from being forgotten by posterity by having the good fortune of being attacked in print by the more talented John Locke. Filmer is a little-known figure today, but he was more influential in the seventeenth century, at least among royalists for whom he provided ideological cover. Book One of Locke's *Two Treatises of Government* is a lengthy dissection of Filmer's *Patriarcha*, which is based on his reading of the biblical figure of Adam. Modern readers of Locke tend to skip the first book and go right to the second, which contains the most influential statement of Locke's positive political theory, based primarily on the "law of Nature" rather than on scripture, though Locke thought that scripture corresponds to the code of conduct discovered by reason in the natural law. It sometimes goes unrecognized, however, that Filmer too was part of the larger seventeenth-century attempt to do natural theology. Rather than build his theory of kingship on an assortment of biblical texts about kings, as the previous tradition had often done, Filmer attempted to embed his theory of kingship in nature, with Adam as a cipher for the natural condition given to human beings in creation.[47] Filmer's *Patriarcha* is subtitled *or the Natural Power of Kings*. Filmer argues that human kingship is given by God, the author of nature; kingship is an extension of the "natural and private dominion of Adam."[48] Filmer identifies the kingdom with the family, and royal power with the power of the father within the family. According to Filmer, God gave to Adam at his creation dominion over the woman and over the rest of creation. This paternal power, which Filmer simply equates with kingly power, was then transmitted to Adam's posterity, and on down to the kings who ruled in Filmer's time. "This Lordship which *Adam* by Command had over the whole World, and by Right descending from him the *Patriarchs* did enjoy, was as large and ample as the Absolutest Dominion of any *Monarch* which hath been since the Creation."[49]

47. W. S. Carpenter, "Introduction," in John Locke, *Two Treatises of Government* (New York: Dutton, 1978), xi.

48. This is Locke's paraphrase of Robert Filmer in Locke, *Two Treatises* 1.7.73 (p. 52).

49. Robert Filmer, *Patriarcha, or the Natural Power of Kings* (London: Ric. Chiswell, 1680), 13.

Filmer's scheme only works because of the common Christian assumption that Adam stands as a representative of all humankind, and so the sin of Adam becomes the sin of all; the Fall is not just the fall of two people, but the Fall of the descendants of Adam, who now must live with the consequences. The transmittal of kingly power to Adam's posterity depends on Adam's representative role.[50] However, Filmer departs from the tradition because the Fall seems to have little practical effect on his political theory. Filmer refutes the idea that humans have what Locke calls "natural freedom" not by contending that they lost that freedom because of human sin, but by contending that all were subject to Adam from the point of Adam's creation. The Fall in Filmer is rebellion against that subjection—he contends that "the desire of Liberty was the first Cause of the Fall of *Adam*"[51]—but the natural necessity of subjection to Adam and his heirs is the same pre- or post-Fall. It is not simply that, like Aquinas, Filmer presses the origin of government back to before the Fall; rather, in discontinuity with the previous tradition of Christian political thought, the Fall has no effect at all. Locke points this out when he criticizes Filmer for contradicting himself. Filmer says, on the one hand, that Adam was monarch as soon as he was created, given absolute power over the whole earth in Genesis 1:28, and on the other hand, that Genesis 3:16—in which the woman is subjected to the man—is the "original grant of government."[52] For Filmer, the discrepancy does not seem to be a contradiction precisely because the Fall that intervenes between 1:28 and 3:16 has no effect worth mentioning.

Locke, however, points out that Genesis 1:28 does not give Adam dominion over other humans,[53] an observation that seems to align Locke with Augustine on this point. Locke also points out that Genesis 3:16 is a curse on both the man and the woman, not the granting of a special monarchical dignity to Adam. As Locke tartly observes, "it would be hard to imagine that God, in the same breath, should make [Adam] universal monarch over all mankind, and a day-labourer for his life. Turn him out of Paradise 'to till the ground,' and at the same time advance him to a throne and all the privileges and ease of absolute power."[54]

Locke seems here to be rescuing the importance of the doctrine of the Fall for political theory. But he is only getting warmed up in his critique of

50. Ian Harris, *The Mind of John Locke: A Study of Political Theory in Its Intellectual Setting* (Cambridge: Cambridge University Press, 1994), 233.

51. Filmer, *Patriarcha*, 3. This is the only reference to the Fall in Filmer's work.

52. Locke, *Two Treatises* 1.3.16 (p. 13).

53. Locke, *Two Treatises* 1.4.26 (p. 19).

54. Locke, *Two Treatises* 1.5.44 (p. 32).

Filmer. For Locke, the most significant vulnerability of Filmer's theory is in the idea of Adam's representation of all humankind. Because Filmer's theory depends on Adam's representative status, Locke attacks precisely this notion, and in so doing, severely curtails the importance of the Fall for humanity following Adam. Locke criticizes Filmer for interpreting what was spoken only to Adam in 3:17ff. as a curse applied to all humankind.[55] Similarly, Locke questions the idea that the curse laid upon Eve should then become a natural law binding on all subsequent women. According to Locke,

> God in this text gives not, that I see, any authority to Adam over Eve, or men over their wives, but only foretells what should be the woman's lot, how by His Providence he would order it so that she should be subject to her husband, as we see that generally the laws of mankind and customs of nations have ordered it so, and there is, I grant, a foundation in Nature for it.[56]

Locke curtails the effects of the Fall; the subjection of women to their husbands is conventional, based in human custom which is generally guided by God's providence, but which can change depending on each individual woman's "condition or contract with her husband." Such subjection somehow also has "a foundation in Nature"; whether this Nature is pre- or postlapsarian does not seem to matter to Locke. What matters is refuting Filmer's use of the common Christian notion that Adam (and Eve) can represent all of humankind.

In *The Reasonableness of Christianity*, Locke rejects those who "would have all *Adam's* Posterity doomed to Eternal Infinite Punishment for the Transgression of *Adam*, whom Millions had never heard of, and no one had authorized to transact for him, or be his Representative."[57] The idea that no one can be represented by another without his or her authorization is certainly not just a theological concept, but has great political purchase for Locke. His theory of political authority places great emphasis on the natural freedom into which each human being is born, a freedom that includes the liberty of each individual agent to, in some sense, choose or consent to his or her own representative. Locke further disfavors the idea of original sin precisely because he rejected the notion that a person can bind his posterity in any and all matters political.

---

55. Locke, *Two Treatises* 1.5.46 (pp. 32–33).
56. Locke, *Two Treatises* 1.5.46 (pp. 33–34).
57. John Locke, *The Reasonableness of Christianity, as Delivered in the Scriptures*, in *John Locke: Writings on Religion*, ed. Victor Nuovo (Oxford: Clarendon, 2002), 91.

It is true that whatever engagements or promises any one made for himself, he is under the obligation of them, but cannot by any compact whatsoever bind his children or posterity. For his son, when a man, being altogether as free as the father, any act of the father can no more give away the liberty of the son than it can of anybody else.[58]

In similar fashion Locke denies that the children of slaves are also born into slavery.[59] In thus undercutting the mechanism by which God's grant to Adam was passed on to Adam's posterity, Locke undermines one of the foundations on which Filmer's theory rests.

Locke does not deny the reality of the Fall. References to the Fall are sprinkled throughout his writings, and he wrote two very short pieces on the subject, one titled *Peccatum originale* from 1692 and the other *Homo ante et post lapsum*, dated a year later. In the former, Locke rejects not the Fall as such but the imputation of Adam's sin to his posterity. Locke argues that it is unreasonable to say that someone really can "participate with Adam in that sin who did not concurr to it by any act of theirs nor were in being when 'twas committed."[60] Locke furthermore rejects the idea that God—while not regarding us as having sinned in Adam—nevertheless subjects us to the evils which were due to Adam as punishment for committing the sin. Locke rejects the first option as impugning God's veracity and the second as impugning God's justice.[61] In *Homo ante et post lapsum*, Locke claims that the original humans were created mortal but would have been granted immortality had they passed the test provided by the "probationary law," that is, the commandment not to eat from the tree of life. They failed the test, and so all now know death. For Adam and Eve, death was punishment, but for their posterity it is simply a consequence of being a corporeal being. "This was the punishment of that 1st sin to Adam & Eve. viz death & the consequence but not punishment of it to all their posterity for they never haveing had any hopes or expectation given them of immortalitie, to be borne mortal as man was first made cannot be called a punishment."[62] Mortality is a natural condition that is not altered by Adam and Eve's sin; what was removed was only the opportunity for immortality. Locke makes the same argument at greater length a few years later (1695) in *The Reasonableness of Christianity*,

58. Locke, *Two Treatises* 2.8.116 (p. 176).
59. Locke, *Two Treatises* 2.16.189 (p. 214).
60. John Locke, *Peccatum originale*, in *John Locke: Writings on Religion*, 229.
61. Locke, *Peccatum originale*, 229-30.
62. John Locke, *Homo ante et post lapsum*, in *John Locke: Writings on Religion*, 231.

in which it is not guilt but only death that "came on all Men by *Adam's* sin."[63] Locke, like Hobbes, thus strives to limit the effects of the Fall to mortality, a mortality that is part of the natural condition of human beings before the Fall, but is merely not removed because of the Fall.

When we look at the architecture of his positive account of the origins of political authority in the second *Treatise*, we see that Locke's "state of nature" depends on an eclipse of the importance of the Fall for political theory. As in Hobbes, Locke constructs a hypothetical state of nature to justify political authority. Against Hobbes, however, the state of nature is not a state of war, nor is it a state of subjection, as in Filmer. Locke writes,

> To understand political power aright, and derive it from its original, we
> must consider what estate all men are naturally in, and that is, a state of
> perfect freedom to order their actions, and dispose of their possessions
> and persons as they think fit, within the bounds of the law of Nature,
> without asking leave or depending on the will of any other man.[64]

The state of nature is also a state of equality, people being "promiscuously born to all the same advantages of Nature . . . without subordination or subjection,"[65] unless God explicitly arranged it otherwise. Whether or not this condition is pre- or post-Fall is hard to judge; in the second *Treatise* all talk of the Fall simply vanishes. The prelapsarian/postlapsarian axis so crucial to patristic and medieval political theory gives way to pre-/post-state of nature as the critical dividing line.

When Locke writes that the state of nature and the state of war "are as far distant as a state of peace, goodwill, mutual assistance, and preservation; and a state of enmity, malice, violence and mutual destruction are one from another,"[66] one would suppose that the state of nature is a prelapsarian condition. But Locke writes that, in the state of nature, the execution of the law of nature that governs in that condition is left to each individual; each man punishes transgressions against himself, "but only to retribute to him so far as calm reason and conscience dictate, what is proportionate to his transgression."[67] The state of nature, then, would appear to be postlapsarian, since there are transgressions there, but if so, "calm

---

63. Locke, *Reasonableness*, 92.
64. Locke, *Two Treatises* 2.2.4 (p. 118).
65. Locke, *Two Treatises* 2.2.4 (p. 118).
66. Locke, *Two Treatises* 2.3.19 (p. 126).
67. Locke, *Two Treatises* 2.2.8 (p. 120).

reason" has not been much damaged by the Fall. Reason, which Locke equates with the law of nature,[68] has been given by God for the mutual security of human persons.[69]

Locke is explicit that both natural reason and scripture affirm that God has given the earth to all in common.[70] How then does Locke explain the fact of private property and inequality in its possession? Not by means of a Fall. It is reason, says Locke—and not sin—that causes the world to be divided up: "God, who hath given the world to men in common, hath also given them reason to make use of it to the best advantage of life and convenience."[71] Locke famously derives private property from labor; when I pick an apple, I "mix" my labor with it to enclose it from the commons. Each person has a "'property' in his own 'person,'" such that anything he removes by his own labor from the state of nature becomes his own property, to the exclusion of others.[72] This applies even to "the turfs my servant has cut" which "become my property without the assignation or consent of anybody,"[73] despite the obvious fact that my servant has labored, not me. Locke also does not explain why labor is inherently private, that is, why common labor is unthinkable.

The crucial point here is that the Fall simply does not apply. Locke derives the right of private property through labor from God's command to subdue the earth—which comes in Genesis 1:28, pre-Fall—but combines it with God's command to labor and till the earth, which is—as Locke has already pointed out in his polemic against Filmer—a post-Fall curse, given to the man as God's punishment in Genesis 3:17ff.

> God, when He gave the world in common to all mankind, commanded man also to labour, and the penury of his condition required it of him. God and his reason commanded him to subdue the earth—*i.e.*, improve it for the benefit of life and therein lay out something upon it that was his own, his labour. He that, in obedience to this command of God, subdued, tilled, and sowed any part of it, thereby annexed to it something that was

---

68. "The state of Nature has a law of Nature to govern it, which obliges every one, and reason, which is that law, teaches all mankind who will but consult it, that being all equal and independent, no one ought to harm another in his life, health, liberty, or possessions"; Locke, *Two Treatises* 2.2.6 (p. 119).

69. Locke, *Two Treatises* 2.2.8 (pp. 120–21).

70. Locke, *Two Treatises* 2.5.25 (p. 129).

71. Locke, *Two Treatises* 2.5.26 (p. 129).

72. Locke, *Two Treatises* 2.5.27 (p. 130).

73. Locke, *Two Treatises* 2.5.28 (p. 130).

his property, which another had no title to, nor could without injury take from him.[74]

Thus does Locke combine two passages from Genesis—one pre-Fall, one post-Fall—into a seamless argument for what is the "natural" condition of humankind.[75] Unremitting toil, inequality, and the enclosure of the commons is not a symptom of the Fall, but simply the way that God and Nature have arranged to make best use of the creation.[76]

Locke continues on to say that because God wanted humans to receive the benefits of creation, God never intended property to remain common; "He gave it to the use of the industrious and rational (and labour was to be his title to it); not to the fancy or covetousness of the quarrelsome and contentious."[77] Different degrees of industriousness result in different proportions of possessions;[78] the native nations of North America have little, despite the abundance of nature, because they have done little by way of "improving it by labour."[79] In the state of nature, people can only have as much as they could make use of; there is no point in gathering more apples than one could use, because they quickly spoil, so the excess apples belong by right to others.[80] The invention of money, however, makes possible great accumulations of wealth, beyond what one can use, because it turns the perishable into the imperishable.[81] Though Locke here has eschewed any role for the Fall or original sin, his account of primitive accumulation corresponds to what Karl Marx calls the "history of economic original sin." Just as the story of Adam explains human misfortune through human sin, says Marx, so goes the story preferred by political economists in accounting for inequality:

> Long, long ago there were two sorts of people; one, the diligent, intelligent and above all frugal elite; the other, lazy rascals, spending their sub-

74. Locke, *Two Treatises* 2.5.32 (p. 132). Nature is indifferent for Locke, not hostile, as in Hobbes.

75. For an astute commentary on this point, see Roland Boer, "John Locke, the Fall, and the Origin Myth of Capitalism," *Political Theology* blog, 5 December 2013, http://www.politicaltheology.com/blog/john-locke-the-fall-and-the-origin-myth-of-capitalism/.

76. To be clear, Genesis does not regard all labor as such to be an effect of the Fall; Gen. 2:15 gives the land, pre-Fall, to the man to cultivate and care for it.

77. Locke, *Two Treatises* 2.5.34 (pp. 132–33).

78. Locke, *Two Treatises* 2.5.48 (p. 140).

79. Locke, *Two Treatises* 2.5.41 (p. 136).

80. Locke, *Two Treatises* 2.5.31 (p. 131).

81. Locke, *Two Treatises* 2.5.36 (p. 134) and 2.5.45–50 (pp. 139–41).

stance, and more, in riotous living. . . . And from this original sin dates the poverty of the great majority who, despite all their labour, have up to now nothing to sell but themselves, and the wealth of the few that increases constantly, although they have long ceased to work.[82]

Locke tells a version of this same tall tale, but for Locke the theological notion of original sin has no traction. Inequality of wealth is not a punishment resulting from the Fall, but a product of mutual human consent: "But since gold and silver, being little useful to the life of man, in proportion to food, raiment, and carriage, has its value only from the consent of men . . . it is plain that the consent of men have agreed to a disproportionate and unequal possession of the earth."[83] Although Locke is anxious that children be free of their parents' contracts, commitments, and sins, Locke does not seem to object to children inheriting their parents' wealth, though those children might not have labored at all.

To say that the traditional doctrine of the Fall and original sin has little effect on Locke's political theory is not to say that Locke has no sense of the corruption of human nature.[84] Scattered throughout Locke's writings are references to the depravity and corruption of human life. For example, Locke ends his *Homo ante et post lapsum* with an account of how corruption came upon the world after Adam and Eve's sin:

upon their offence they were affraid of god, this gave them frightfull Ideas and apprehensions of him & that lessened their love which turnd their minds to the creature this root of all evill in them made impressions & soe infected their children, & when private possessions & labour which now the curse on the earth had made necessary, by degrees made a distinction of conditions, it gave roome for coviteousnesse pride & ambition, which by fashen & example spread the corruption which has soe prevailed over man kind.[85]

Locke shares with his contemporaries a keen sense of corruption in the world, but here he gives a historical and social rather than ontological account of its origins. Corruption spread by "fashen & example," not by biological reproduction or any other mechanism that would alter human being.

82. Karl Marx, *Capital*, vol. 1, trans. Ben Fowkes (New York: Vintage, 1976), 873.

83. Locke, *Two Treatises* 2.5.50 (p. 140).

84. For an account of human corruption in Locke, see W. M. Spellman, *John Locke and the Problem of Depravity* (Oxford: Clarendon, 1988).

85. Locke, *Homo ante*, 231.

In an Augustinian framework, concupiscence was an inherited effect of the Fall. In Locke's second *Treatise,* he refers to a "Golden Age" "before vain ambition, and *amor sceleratus habendi,* evil concupiscence, had corrupted men's minds."[86] But this time appears to be postlapsarian, because there is government. What this suggests is that corruption is not inherent in human being, but an effect of social causes. Ian Harris concludes that "[i]t seems hard to deny that if human capacities were impaired by the Fall it was to a degree hardly worth mentioning for Locke's purposes."[87] Locke's educational works, Harris notes, attribute human biases to nature, not to the Fall. Peter Harrison similarly concludes that for Locke the limits to human capacities are not so much the result of the Fall, but rather of the natural limits inherent in being embodied creatures created in a middle position between angels and beasts. In this condition, we are little different from Adam.[88]

For Locke, as for Hobbes and Filmer, political theory based in the Fall was replaced by political theory based in nature. For Locke, God's intentions for creation show through in both scripture and nature. "Nature teaches all things,"[89] says Locke; "God," "nature," and "reason" are terms that Locke uses almost interchangeably as sources of human knowledge.[90] Scripture, properly interpreted, confirms the natural law. The second table of the Decalogue, for example, corresponds to the code of conduct discoverable by reason.[91] Scripture is not simply superfluous for Locke; it was still needed to cover for the weaknesses in human reason. But we see in Locke that scripture has lost much of its load-bearing role, and political theory rests on an account of the state of nature and the social contract by which we remedy the deficiencies of nature, not on any account of life pre- and post-Fall.

Conclusion

Hobbes and Locke are rightly considered founding fathers of modern political theory because of their attempts to build a new political "science" on

86. Locke, *Two Treatises* 2.8.111 (p. 173).

87. Harris, *The Mind of John Locke,* 299.

88. Peter Harrison, *The Fall of Man and the Foundations of Science* (Cambridge: Cambridge University Press, 2007), 223–34, 232.

89. Locke, *Two Treatises* 1.6.56 (p. 40).

90. "God and Nature" appear frequently together as endowing humans with certain capacities; for example, Locke, *Two Treatises* 1.9.90 (p. 63).

91. Harris, *The Mind of John Locke,* 31–32.

a more naturalistic basis. Both Hobbes and Locke used scripture to lend its contested authority to more fundamental laws of nature that could be discovered by human reason. They are considered the first modern rather than the last medieval political theorists because both paved the way for the public authority of the Bible to be supplemented, and eventually replaced, by Nature as the secure foundation of knowledge.

The eclipse of the Fall in early modern political theory coincides with a new, unitary conception of nature. As we have seen, the Fall marks a division between two kinds of nature, the way we are and the way we are meant to be. The Fall is therefore crucial to an eschatological concept of nature; nature is not simply there, inert, its constant properties to be investigated and codified into constant laws. The Fall marks the fact that nature has a goal, a *telos*; there is nature as it is and nature as it will become, the latter of which is revealed by reflection on the original, prelapsarian condition in which God placed us, which in turn reveals God's intention for us. Modern science rejects teleology, believing the nature of matter to include only the way things are, and not the way things ought to be. The new "science" of politics also collapses the two natures into one; the way things are is revealed by the state of nature, which politics can ameliorate but not essentially alter. The Fall is "naturalized," and many of the features of fallenness now simply coincide with creaturehood.

In this chapter, however, I hope to have cast doubt on the inevitability of this process of naturalization. The eclipse of the biblical Fall story was not simply the putting away of childish stories in favor of hard data; the eclipse of the Fall was at least in part political, not scientific. The fall of the Fall is part of the secularization of politics, but secularization is neither inevitable nor the simple subtraction of a supernatural worldview from some more basic, natural residue. The "state of nature" upon which Hobbes and Locke built their political theories is based not on any empirical testing, but rather on prior political decisions about what kind of government and political economy needs justification. Marx was right to accuse Locke and others of substituting one original sin story for another. The state of nature replaced the Fall with a story of human origins that is no more empirically based and no less susceptible to being labeled "mythological" than the Genesis story. The claim to know "nature" and what is "natural" is, for Hobbes and Locke, a political move that, no less than medieval appeals to scripture, attempts to invest politics with authority that comes from a nonpolitical source.

The story of early modern political theory is a story of secularization, but not in the sense that the term is usually used today. God was not bracketed out—early modern European states vociferously claimed direct divine authority—but such authority was increasingly unmediated by the church, which in many cases was reduced practically to an office of the state. The original meaning of the term "secularization" was the transfer of property or power from ecclesiastical to civil control.[92] In this sense, both Hobbes and Locke contributed to the secularization of politics, which was not yet (if it ever has been) a desacralization.

The eclipse of the biblical Fall in particular had significant advantages for the justification of the authority of the nascent modern state, which was busily freeing itself from ecclesiastical interference and appropriating land, judicial powers, rights of appointment to ecclesiastical offices and benefices, and tax powers and revenues from the church. The movement of politics from a scriptural to a "natural" basis meant less reliance on the church for its expertise in biblical interpretation. More importantly, the eclipse of the Fall removes the eschatological proviso that the medieval commentators read in the Genesis story. For Augustine and the tradition that followed him, the Fall meant that coercive political authority was not natural or permanent but a temporary remedy for sin until the return of Christ, the true ruler. Political authority, though instituted by God, lived always under the judgment of the way things were meant to be, which was, of course, also God's judgment. In contrast, the state that emerges from the "state of nature" is simply a response to the way things are, and therefore a natural, permanent, institution.

In the long run, much of what comes to be called "science" will follow the path that "political science" followed: divorced from theology and from the church, and tasked with investigating a reduced nature that has been stripped of any eschatological or teleological reference, the Fall will be discarded as a quaint myth, and evolution will appear to be guided by purely immanent processes. I have suggested, however, that the divorce of science and theology in the West has been promoted, at least in part, by nonscientific factors. I have offered this examination of the politics of the Fall as a contribution to a political history of science in the West. If this history is correct, then perhaps antagonism between science and theology is not inevitable, and perhaps it is possible to give an account of evolution and the Fall that is true to both the scientific evidence and Christian revelation.

92. Jan N. Bremmer, "Secularization: Notes toward a Genealogy," in *Religion: Beyond a Concept*, ed. Hent de Vries (New York: Fordham University Press, 2008), 432–33.

Augustinian Reflections on Christianity and Evolution

PETER HARRISON

In educated circles, religiously motivated resistance to science has a bad name. Its most conspicuous contemporary manifestation, young earth creationism, is associated with an undesirable religious fundamentalism, right-wing politics, bigotry, and backwardness. While the responses of critics of religion are predictable—"ignorant, stupid or insane" is Richard Dawkins's dismissive characterization of anti-evolutionism—more significant is the fact that mainstream Christian denominations take a similarly dim view of scientific creationism, even if their language is more moderate.[1] If scientific creationism represents the present face of the religious repudiation of science, the best-known past instance of conflict is the notorious 1633 condemnation of Galileo by the Holy Office. For the Catholic Church this episode is now seen as a tragic misjudgment. For others sympathetic to religion (and historical accuracy, for that matter) this is an event that calls for careful historical analysis in order to establish that this was not just about science and religion, and that it was by no means typical of a Catholic attitude toward science.

These two prominent instances of science-religion conflict play a central role in virtually all discussions about science and religion. Those dismissive

---

1. Richard Dawkins, "Put Your Money on Evolution," *New York Times Review of Books*, April 9, 1999, 35. For mainstream religious responses, see, e.g., statements by over twenty religious organizations on the National Center for Science Education (NCSE) website: http://ncse.com/media/voices/religion, accessed May 25, 2014. For a slightly more nuanced Catholic response, see Pius XII, *Humani generis* (1950), http://www.vatican.va/holy_father/pius_xii/encyclicals/documents/hf_p-xii_enc_12081950_humani-generis_en.html, accessed May 25, 2014.

of religion regard them as emblematic of the irrationalism of religious faith and enlist them as part of a general critique of religion. For their part, those in mainstream religious traditions find these two examples equally instructive, taking them to support the generalization that conflict between science and religion is almost always undesirable. Such conflict is perceived to have the potential to undermine the credibility of religion and give succor to its opponents.

In this chapter I will consider this latter position, and suggest that the advocacy of peaceful relations between science and religion that we encounter in mainstream religious groups arises, in part, from an absence of instances of what we might call "good conflicts" or "justifiable conflicts." Creation science and the Galileo affair offer examples in which the relevant science, either in prospect or retrospect, seems undeniably correct. The Galileo affair, in particular, is deployed time and time again to exemplify the folly of religious opposition to science, and often in the context of illustrating the folly of resistance to evolution. But what if there were other examples that were less clear-cut, and that might offer alternative models of creative tension or outright conflict? This chapter will consider some possible candidates for such a role, and offer some tentative conclusions about what might follow for the issue of the pressures placed on traditional doctrines of human origins and the origins of sin by the theory of evolution. We will do well to remember the wisdom of Augustine: "The obscure mysteries of the natural order, which we perceive to have been made by God the almighty craftsman, should rather be discussed by asking questions than by making affirmations."[2]

Two Models of Peaceful Interaction

At the outset it is worth making a few preliminary and clarifying remarks. The term "mainstream religious groups," used above, may seem somewhat tendentious, so it is worth specifying briefly the constituency that I am seeking to characterize here. This label takes in contemporary Catholicism, Anglicanism, Lutheranism, and most Reformed churches. Also characteristic of the irenic position that I wish to explore are major organizations such as the Templeton Foundation, BioLogos, and the International Society for Science and Religion, which sponsor research and activities that promote friendly

2. Augustine, *De Genesi ad litteram imperfectus liber* 1.1.

relations between science and religion. To take a single example, BioLogos advocates "harmony between science and faith" along with "an evolutionary understanding of God's creation."[3] Finally, there are a number of academic journals devoted to the discussion of science-religion relations that focus mostly on constructive dialogue.[4] These groups have a recognizable identity in science-religion discussions and have been labeled, often in pejorative terms, as "accommodationists" or "neo-harmonizers."[5]

Among these groups that I take to be the foremost promoters of peaceful relations between science and religion are two approaches to the issue. These approaches are sometimes implicit, and while they represent two quite distinct positions they are at times conflated. One position, the "hard" irenic position, is that conflict between science and religion is, *in principle*, not possible. This could be because science and religion are regarded as dealing with independent spheres. Since their interests do not overlap, conflict cannot occur. More often, though, advocates of the hard irenic position appeal to the principle, articulated by Pope Leo XIII, that "truth cannot contradict truth": veridical science and true religion should in principle never come into conflict.[6] This assumes that both science and religion are in some sense truth-tracking. Closely related to this principle is the longstanding motif of the "two books"—the book of nature and the book of scripture—both of which are authored by God. Again, the notion is that since they share the same divine author, the Bible and the study of nature cannot come into conflict. The two books metaphor can be found as far back as Augustine of Hippo (354–430), was commonplace during the Middle Ages, and was elaborated by key figures of the seventeenth-century scientific revolution such as Francis Bacon and Galileo. When applied to specific cases, the hard irenic position suggests that because there can be no genuine conflicts between sci-

---

3. http://biologos.org/about, accessed May 23, 2014.

4. The journals I have in mind are *Zygon, Theology and Science, Science and Christian Belief, Perspectives on Science and Christian Faith*, and *Philosophy, Theology and the Sciences*.

5. For the former see, e.g., Jerry Coyne, "Accommodationism and the Nature of Our World," http://whyevolutionistrue.wordpress.com/2009/04/30/accommodationism-and -the-nature-of-our-world/; for the latter, David A. Hollinger, *After Cloven Tongues of Fire: Protestant Liberalism in Modern American History* (Princeton: Princeton University Press, 2013), 82–102. Also see Peter Harrison, "The Neo-Harmonists: Rodney Stark, Denis Alexander, and Francis Collins," in *The Idea That Wouldn't Die: The Warfare between Science and Religion* (Baltimore: Johns Hopkins University Press, 2017).

6. For the principle that truth cannot contradict truth see Leo XIII, Encyclical *Providentissimus Deus*. http://w2.vatican.va/content/leo-xiii/en/encyclicals/documents/hf_l -xiii_enc_18111893_providentissimus-deus.html, accessed July 4, 2015.

ence and religion, putative tensions can be resolved by showing that science or religion has made claims that lie beyond its respective sphere, or that it has not been true to its mission in other ways. In practice, though, given the present high status of science, this usually means modifying religious claims rather than seeking to adjust the relevant science.

The alternative view, the "soft" irenic position, holds that concord between science and religion is much more a matter of historical contingency. Peace is a good thing, but it occurs because at that particular time the relevant science just happens not to conflict with religion. It might be the case, for example, that evolutionary theory does not conflict with a Christian view of creation. But for advocates of the soft irenic position, this position is not derived from any overarching principle about necessary relations between science and religion, and no generalization about science-religion relations will follow from it. It is just that examination of the relevant scientific and religious doctrines yields, in this particular case, no evidence of conflict. So while advocates of the soft irenic view might argue that nothing in contemporary science need be of particular concern to believing Christians, this would not rule out potential conflict in the future, nor the possibility of genuine conflicts in the past. In short, both irenic positions share the view that genuine conflict between science and religion is *never inevitable*, but they differ on whether it is possible at all.

This soft irenic stance involves a much less essentialist understanding of science and religion which, in the hard view, are taken always to proceed along certain lines such that their ideal relationship is independent of particular historical manifestations of the relevant activities. (Proponents of inevitable conflict hold a similarly essentialist view, contending that science is always based on reason and experience, and religion on faith and authority.) Alternatively, a possibility for advocates of the soft irenic position is the claim that science is not consistently truth-tracking—a view that could be supported by pointing to the fact that considered diachronically, scientific claims made at one time have conflicted with those made at another. In neither case is there an overarching framework that determines the legitimate scope of each enterprise in a way that would definitively rule out the possibility of conflict.[7]

---

7. Soft irenicism might also question whether there are enduring entities "science" and "religion." For an account of the changing historical understandings of science and religion see Peter Harrison, *The Territories of Science and Religion* (Chicago: University of Chicago Press, 2015).

My suggestion in this chapter will be that the soft irenic position has significant and unappreciated merits. One of its implications is that potential science-religion conflicts need to be considered on a case-by-case basis. This stance also prompts us to look closely at the details of various scientific claims, with the possibility that some aspects of general theory might be acceptable, but others not. Specifically, in the case of evolutionary theory the argument would *not* be that there is a scientific consensus about the truth of evolution and that therefore Christian thinking must adapt itself to this reality. Rather, it would be a matter of scrutinizing every element of the theory, its variant forms and their implications, and considering whether all or some or none were compatible with core Christian beliefs. There is, of course, the related issue of what counts as "core Christian belief," but for the moment we are simply speaking in the abstract.

In order to explore this issue further I want to turn to Augustine's way of dealing with tension between scientific doctrines about the natural world and Christian teachings. Augustine often makes an appearance in science-religion discussions as an example to be emulated. He is, of course, a thinker of enormous stature, arguably the most significant Christian author outside the New Testament writers. Augustine developed highly influential views about the origins of human beings and of sin. He also articulated a number of sophisticated principles for dealing with the relations between Greek science and Christian thought—principles that were later put to use by Galileo in defense of his own cosmological views.[8] Augustine is also touted as a precocious evolutionist on account of his appeal to developmental principles in the organic realm. Accordingly, his thinking is relevant to the two most prominent cases of science-religion conflict: evolution and the Galileo affair. In both cases, he is typically regarded as exemplifying a hard irenicism.[9] My suggestion will be that Augustine does indeed offer an exemplary model of

8. Ernan McMullin, "Galileo on Science and Scripture," in *The Cambridge Companion to Galileo*, ed. Peter Machamer (Cambridge: Cambridge University Press, 1999), 271–347. Pietro Redondi links Galileo and Augustine in another way, suggesting that Galileo's Augustinian theological commitments were central to his mechanics. "From Galileo to Augustine," in *The Cambridge Companion to Galileo*, 175–210.

9. Thus, McMullin: "He presupposes, of course, as a first principle that no *real* conflict can arise"; "Galileo on Science and Scripture," 291. Kenneth Howell speaks similarly of Augustine's insistence that there can be no contradiction between "biblical truth and true knowledge from outside the Bible." See Howell, "Natural Knowledge and Textual Meaning in Augustine's Interpretation of Genesis," in *Nature and Scripture in the Abrahamic Religions: Up to 1700*, ed. Jitse van der Meer and Scott Mandelbrote, 2 vols. (Leiden: Brill, 2008), 2:117–46 (141).

dealing with apparent science-religion conflicts but that he is not, as often claimed, an advocate of hard irenicism. Moreover, he offers examples of fruitful conflict that have a bearing on contemporary discussions of evolution in relation to theological anthropology.

Augustine, Natural Science, and Creation

Augustine's generally positive attitude toward pagan thought is typically juxtaposed to that of another church father, Tertullian. The latter famously contrasted Athens and Jerusalem, expressed scorn for the "wisdom of the world," and his caricatured (and misquoted) *credo quia absurdum*—"I believe because it is absurd," is taken to be emblematic of the conflict model.[10] Philosophy, Tertullian wrote, was evasive, farfetched, harsh, productive of contentions, and embarrassing even to itself. Moreover, it was the source of heresy.[11] (In the ancient, medieval, and early modern worlds, "natural philosophy" or often simply "philosophy" was the enterprise that most resembled modern science. "Science," in Anglophone contexts, really only takes on its present meaning in the nineteenth century.) Not surprisingly, then, Tertullian almost invariably makes a cameo appearance in works arguing that conflict represents the essential and perennial relationship between science and religion.[12]

Augustine, by way of contrast, is thought of as having enunciated a number of insightful principles for dealing with the potential conflicts between Christianity and pagan natural philosophy. Much of the relevant discussion comes in his treatise *De genesi ad litteram* (On the Literal Interpretation of Genesis), in which he considers a number of problems related to his understanding of creation. These were prompted, in part, by objections of the Manicheans to the biblical accounts of creation. It is important to understand that Augustine deploys the descriptor "literal" here in a way somewhat unfamiliar to modern readers, using it to distinguish his approach in this

10. The phrase *credo quia absurdum* does not appear in Tertullian. The relevant phrases are *"prorsus credibile est, quia ineptum est . . . certum est, quia impossibile"*; *De carne christi* 5. 4 (Patrologia Latina 2:761). See Robert D. Sider, *"Credo quia Absurdum?,"* *The Classical World* 73 (1980): 417–19.

11. Tertullian, *Adversus haereticos* 7.

12. For the polemical uses of Tertullian and the invention of the phrase "I believe because it is impossible," see Peter Harrison, *"Credo quia absurdum*: The Enlightenment Invention of Tertullian's Credo," forthcoming.

work from highly allegorical or moral readings of Genesis that were common during this period.[13] Augustine sought, in his literal meaning, to establish the sense intended by the author. In instances where the author's intended meaning was obscure, a range of interpretations was possible, provided that they were consistent with "sound faith."[14]

One of the central problems for Augustine was an apparent inconsistency among various biblical sources. The first chapter of Genesis suggests a six-day creation, while Genesis 2:4 implies that the world was created in a day. Also at odds with six-day creation was the Apocryphal work, *Ecclesiasticus* (*Sirach*), which also seemed to teach that God had created everything at once (*creavit omnia simul*, Vulgate, 18:1).[15] Furthermore, both of these teachings were difficult to reconcile with ongoing creative processes that saw new things coming to be and passing away in the natural world. Augustine adopts two strategies in relation to this problem. First, he points out that a strictly literal reading of the days of Genesis is not warranted by the text. For a start, the sun was not created until the fourth "day" (as Manichean critics had also pointed out). Since time is measured in terms of the motions of the celestial bodies, the preceding days cannot be twenty-four-hour periods. Augustine reasons that the "days" of Genesis 1 are part of a narrative that has been carefully constructed with the limitations of human understanding in mind, and concludes that God created everything at once.[16]

Augustine's second strategy was to introduce a philosophical (or "scientific") principle derived from Stoic thought, in order to resolve the tension between ideas of complete and instantaneous creation, and empirical evidence of a world of constant change and development. Augustine appeals to the Stoic notion of "seminal principles"—in-built seeds, the presence of which will bring about some future event. Here the idea is that God, in a single creative act, implanted the seeds of everything that was to unfold in the future. In Augustine's words, when God created all things simultaneously, he implanted "seeds in the sense of future realities, destined to germinate in

13. Augustine, *Retractationes* 2.24.1.

14. Augustine, *De Gen. ad lit.* 1.21.41. *On Genesis*, trans. Edmund Hill, in *The Works of Saint Augustine* (New York: New City, 2002), part 1, vol. 13, 188. For helpful overviews of Augustine's views on science and the Bible, to which I am indebted in what follows, see Howell, "Natural Knowledge and Textual Meaning in Augustine," and McMullin, "Galileo on Science and Scripture."

15. *De Gen. ad lit.* 4.33.51. Based on a Latin mistranslation of the Septuagint. Patrologia Latina 34:318.

16. *De Gen. ad lit.* 6.16.27; 4.21.38.

suitable places from hidden obscurity into the manifest light of day through the course of the ages."[17] So living things were created in potential in a single event, to be "brought forth" at a later time. It is on this basis that some have claimed Augustine as an evolutionary thinker.[18]

Another issue for Augustine was the apparent conflict between biblical statements and well-established philosophical or scientific claims. Psalm 104:2, one of Augustine's favorite passages, speaks of the heavens being stretched out like a parchment or a skin. Other biblical passages speak of the "vault" of heaven.[19] Not only did these passages seem to contradict each other, they also conflicted with the scientific consensus that the heavens were spherical.[20] Here Augustine suggests that the issue of the shape of the heavens is not of immediate consequence for "those wishing to learn about the blessed life." For this reason, biblical authors had not been preoccupied with such questions. Moreover, such discussions concern matters that are "obscure" and "far removed from our eyes and experience."[21] But Augustine goes on to suggest that in the case of scientific truths that are supported by "proofs that cannot be doubted" or by "true reasoning," these truths must always take precedence over the literal sense of scripture.[22] Specifically in this instance, spherical heavens were to be preferred to other conceptions. Augustine also considers the case where neither the relevant science nor scripture were entirely clear. Responding to the issue of whether or not the heavenly bodies were moved by intelligences (which was the standard philosophical view at the time) Augustine counsels caution, arguing for "that restraint that is proper to a devout and serious person." On obscure questions it is best not to be overcommitted to any prevailing doctrine, since the truth "may later be revealed."[23]

---

17. *De Gen. ad lit.* 4.11.18 (p. 311); cf. 6.10.17. For a patristic precedent see Gregory of Nyssa, *In Hexameron* (Patrologia Graeca 44:72). Cf. Plotinus, *Enneads* 2.3.16. For Augustine's use of this idea see Étienne Gilson, *The Christian Philosophy of Saint Augustine* (London: Victor Gollancz, 1961), 197–209.

18. Ernan McMullin, *Evolution as a Christian Theme* (Waco, TX: Baylor University Press, 2004); John Zahm, *Evolution and Dogma* (Chicago: D. H. McBride, 1896), 283f.

19. Augustine refers to Isaiah 40:22 LXX; the relevant term is *camera*: "*coelum dicitur velut camera esse suspensum.*" Patrologia Latina 34:272.

20. *De Gen. ad lit.* 2.9.21.

21. *De Gen. ad lit.* 1.18.37 (p. 185).

22. Augustine uses the expressions "*ut dubitari inde non debeat*" and "*veris illis rationibus.*"

23. "*ne forte quod postea veritas patefecerit.*" *De Gen. ad lit.* 2.18.38 (Patrologia Latina 34:280).

By the same token, where such demonstrative proofs were lacking, the plain sense of scripture should take precedence over scientific speculations. Indeed, in some instances where the relevant science seems relatively unassailable, Augustine still insisted on the priority of scripture. Augustine takes the example of waters above the firmament (Gen. 1:7), which on the prevailing understanding of the "proper place" of the elements earth, air, fire, and water, was impossible. He then offers some rather hypothetical proposals that might account for this possibility. But then, and notwithstanding its conflict with the standard scientific teaching, he insists that divine revelation must take precedence over "human ingenuity." This is the principle that truth resides in what God reveals rather than "the guesswork of human weakness."[24] These instances, crudely put, amount to conflicts between science and scripture where scripture takes precedence.

Finally, Augustine observed that some of his fellow Christians espoused erroneous scientific doctrines and argued incorrectly that these had been derived from scripture. Augustine observes that this is deplorable since it casts doubt on the credibility of Christianity and on the general reliability of the scriptures when it speaks on other matters:

> Whenever, you see, they catch out some members of the Christian community making mistakes on a subject which they know inside out, and defending their hollow opinions on the authority of our books, on what grounds are they going to trust those books on the resurrection of the dead and the hope of eternal life and the kingdom of heaven . . . ?[25]

Unlike the other "internal" issues noted above, this was an apologetic concern, related to how outsiders might perceive Christianity. That said, it is not difficult to see how such occasions might result from breaches of Augustine's prohibition of premature commitment to ephemeral scientific doctrines.

It is possible to distill all of this discussion into a few general rules. The distinguished philosopher and historian Ernan McMullin helpfully identified several principles, which I enumerate here.[26]

---

24. *De Gen. ad lit.* 2.5.9, 2.9.

25. *De Gen. ad lit.* 1.19.39, trans. in *Works*, part 1, vol. 13, p. 187 (*Patrologia Latina* 34:261).

26. McMullin, "Galileo on Science and Scripture," 292–329. McMullin listed five principles. I have numbered them 1–4 since 2b is a corollary of 2a.

1. The Principle of Prudence: When attempting to interpret a difficult passage of scripture, entertain as many interpretations as possible, particularly in view of the fact that new truths may come to light in future.
2a. The Priority of Demonstration: When a conflict appears between a proven truth about nature and an interpretation of scripture, scripture should be reinterpreted.
2b. The Priority of Scripture: When there is an apparent conflict between scripture and doctrine about the natural world based on reason or sense, where that latter doctrine is not demonstrated, the literal reading of scripture should prevail.
3. The Principle of Accommodation: The words of scripture are adapted to the capacities of its readers.
4. The Principle of Limitation: The primary concern of scripture is salvation, not science.

These Augustinian principles, or some version of them, are often taken by accommodationists to represent "best practice," and Augustine is held in high regard by advocates of peaceful coexistence between science and religion. These principles received an added fillip from Galileo's deployment of them in his defense of Copernicanism.[27] However, I want to suggest a slightly divergent reading of these principles—one that shows Augustine to be somewhat more ambivalent about the status of natural science, and less committed to a hard irenicism than is often assumed.

Demonstrated Truths and the Blessed Life

I want to focus on three of the principles set out above, beginning with the principle of limitation (4). This is the principle that recognizes the divergent aims of science and scripture. When we attend to what Augustine actually says on this issue, it emerges that he does not merely distinguish between the aims of Christianity and natural philosophy: he also insists that the value of the former is significantly greater than that of the latter. In his discussions on the form of the sky, for example, he remarks:

27. For subtle differences on the positions of Augustine and Galileo, see Eileen Reeves, "Augustine and Galileo on Reading the Heavens," *Journal of the History of Ideas* 52 (1991): 563–79.

Many people, you see, have many arguments about these points, which our authors with greater good sense passed over as not holding out the promise of any benefit to those wishing to learn about the blessed life, and, what is worse, as taking up much precious time that should be spent on more salutary matters.[28]

In fact, throughout Augustine's works, the pursuit of questions about nature is routinely described in unflattering terms. In the text on Genesis discussed above, he speaks of "those men who have indulged their idle curiosity by devoting themselves to the study of these things." Elsewhere he suggests that much of the investigation of nature rests upon "the guesswork of human weakness."[29] In his classic autobiographical work, *The Confessions*, Augustine goes further, attributing much investigation of the natural world to the vice of curiosity: "when people study the operations of nature which lie beyond our grasp, when there is no advantage in knowing, and the investigators simply desire knowledge for its own sake." In this context he was to speak of a "form of temptation," "a lust for experimenting and knowing," "a diseased craving," a "vain inquisitiveness dignified with the title of knowledge and science."[30] Compared to the pursuit of virtue, knowledge of nature was of little value.[31]

The general point here is that not only did Augustine observe a clear distinction between the purpose of scripture and the purpose of natural philosophy (as most commentators have stressed); he also made a significant

---

28. *De Gen. ad lit.* 2.9.10 (p. 201).

29. *De Gen. ad lit.* 2.10.23 (p. 203); 2.9.21 (pp. 201, 202). That said, Augustine does find value in knowledge of nature insofar as it assists in the interpretation of scripture, which at times uses analogies drawn from the natural world.

30. Augustine, *Confessions* 10.35, trans. H. Chadwick (Oxford: Oxford University Press, 1991), 211. For Augustine's treatment of curiosity, see Hans Blumenberg, "*Curiositas* and *veritas*: Zur Ideengeschichte von Augustin, Confessiones X 35," *Studia Patristica* 6, Texte und Untersuchungen 81 (1962): 294–302; "Augustin's Anteil an der Geschichte des Begriffs der theoretischen Neugierde," *Revue des Études Augustiniennes* 7 (1961): 35–70. On curiosity more generally see Lorraine Daston, "Curiosity in Early Modern Science," *Word and Image* 11 (1995): 391–404; Peter Harrison, "Curiosity, Forbidden Knowledge, and the Reformation of Natural Philosophy in Early Modern England," *Isis* 92 (2001): 265–90; Paul J. Griffiths, *The Vice of Curiosity: An Essay on Intellectual Appetite* (Winnipeg: Canadian Mennonite University Press, 2006).

31. Augustine, *City of God* 7.34–35, in *Nicene and Post-Nicene Fathers*, First Series, ed. Philip Schaff (Grand Rapids: Eerdmans, 1956), 10 vols. (hereafter *NPNF* 1), 2:141b–2a; *Homilies on the First Epistle of John* 2.13 (*NPNF* 1, 7:474a).

normative judgment about the relative merits of the two activities and of the superiority of the former to the latter (which fewer commentators have stressed). Augustine's writings actually contain a twofold critique of natural philosophy. At one level, he worries about the tendency of those engaged in the study of nature to forget the most important thing: the status of their own souls. To a degree this reflects a recurring motif in the philosophical tradition up to the early modern period. In the fifth century BCE, Socrates sought to divert philosophical discussion away from the physical world toward moral concerns. Much later, in the fourteenth century, inspired by his reading of Augustine's *Confessions*, the leading Renaissance figure Petrarch similarly asked about the use of knowing all about natural things, but "neglecting man's nature, the purpose for which we are born, and whence and where we travel."[32] Critiques of the new orientation of the new sciences of the seventeenth century make the same point: practical inventions may make life more convenient, but they do not promote moral improvement, which is more important.[33] Owing, perhaps, to a set of latent Platonic commitments, Augustine always worried lest an interest in material things lead to a preferential love of the lower over the higher orders of the creation.

The second aspect of Augustine's critique of philosophy follows on from the first. Rightly oriented philosophy shares with Christianity the quest for the blessed life: "the urge for the blessed life [*beata vita*] is common to philosophers and Christians."[34] But only Christianity offers the means to reach that goal. For Augustine, then, sound knowledge of nature is not in any sense parallel, in terms of its worth, to sound religious knowledge. There will always be an unequal partnership between human knowledge and divine knowledge.

If we were to think about how this applies to present discussions, one way of cashing out this principle would be to ask what is at stake if we turn out to be wrong in our commitments in the respective areas. If our science is wrong, that is one thing; if our fundamental moral convictions turn out to

---

32. Petrarch, *De sui ipsius et multorum ignorantia*, quoted in Cassirer et al., *The Renaissance Philosophy of Man* (Chicago: University of Chicago Press, 1948), 58–59. Cf. Augustine, *Confessions* 10.8; *De vera religione* 39.72.

33. Meric Casaubon, *A Letter of Meric Casaubon, D.D. &c. to Peter du Moulin D.D., concerning Natural Experimental Philosophie* (Cambridge, 1669), 5–6. Henry Stubbe, *The Plus Ultra reduced to a Non-Plus* (London, 1670), 13. On this theme see Harrison, *The Territories of Science and Religion*, 128–31.

34. Augustine, Sermon 150.4, in *Works*, part 3, vol. 5, p. 31. See also *The Trinity* 8.7.10; *City of God* 8.3; 18.41; 19.1; *Of True Religion* 2.2. Cf. Athenagoras, *A Plea for the Christians* 7.

be wrong, that is another. Admittedly, it can be more complicated than this, since there are significant overlaps between descriptive and normative—scientific and moral—domains, as in the case of climate change.

The second principle I want to examine more closely is the priority of demonstration (2a) and its corollary, the priority of scripture (2b). Here Augustine proposes that a proven truth about nature must always take precedence over an apparent literal truth of scripture and, conversely, that in cases of conflict where there is no proof for the relevant science, the literal sense of scripture must take priority. It was this principle that Galileo sought to deploy, based on his confident assumption that he possessed a demonstrative proof of the earth's motion (he didn't). These principles require some further explanation. The vocabulary of "proof" and "demonstration" that Augustine relies upon here derives from Aristotle's influential understanding of the nature of science (*epistēmē*), set out in the *Posterior Analytics*. While there is much that could be said about Aristotle's ideal of scientific knowledge, suffice it to say that for Aristotle knowledge is scientific if it is certain knowledge of causes. The certainty of this knowledge arises out of the fact that it is the end product of logical deduction.[35] This is a remarkably high standard, equating to something akin to mathematical proof. Given this, it is not surprising that Augustine enunciated the principle of the priority of demonstrated truths, since to oppose such truths would be tantamount to defying logic.

We might ask, however, whether scientific knowledge ever satisfies these strict criteria. This question became a point of contention during the scientific revolution, which introduced a new standard of scientific knowledge that was based on induction and experiment. While some key figures in the new sciences sought to hold onto older deductive notions of demonstration and proof, others conceded that no form of human knowledge could actually meet the strict criteria originally set out by Aristotle.[36] John Locke (1632–1704), the philosopher most connected with the new experimental sciences of the period, thus declared that "experience may procure us convenience, but not science [in the Aristotelian sense]." Ratio-

35. From an extensive literature on this topic see, e.g., Richard McKirahan Jr., *Principles and Proofs: Aristotle's Theory of Demonstrative Science* (Princeton: Princeton University Press, 1992).

36. Galileo and Descartes, arguably, still clung to some ideal of demonstrative certainty in the sciences. This was also one of the key points at issue between Galileo and Bellarmine. The latter accepted the principle of the priority of demonstration, but denied that Galileo possessed such a demonstration.

nal and regular experiments, he observed, might help us see into the nature of things; however, "this is but judgment and opinion, not knowledge and certainty."[37] Locke, interestingly, continues by endorsing Augustine's priorities: "For 'tis rational to conclude, that our proper Imployment lies in those Enquiries, and in that sort of Knowledge, which is most suited to our natural Capacities, and carries in it our greatest interest, *i.e.*, the Condition of our eternal Estate."[38] For Locke, experimental science (in our sense) may offer us conveniences, but it does not give us any certainties. Moreover, human beings are naturally oriented toward another task for which they are far better equipped—knowledge of God, of ourselves, and of our moral duties. If Locke's understanding of experimental science is correct, then strict application of the Augustinian Principle of the Priority of Demonstration would mean that no science would carry sufficient weight to displace a literal scriptural claim, since no science ever meets the strict criterion of demonstrative certainty.

At this point, while we are considering the status of scientific claims, it is worth introducing a third Augustinian principle—(3) the principle of prudence. This principle advocates keeping an open mind about aligning biblical teachings with current science since "new truths may later appear." Here Augustine offers another relevant observation about scientific claims: they change over time. Less pejoratively, we might say that science makes progress (even if, as theoretical physicist Max Planck wryly observed, it advances one funeral at a time). Granting this, it is important not to be too committed to prevailing scientific speculations since these may later change. Indeed, overcommitment to the status quo by the scientific community constitutes a significant barrier to the uptake of new scientific theories. The principle of the shifting nature of scientific orthodoxy was invoked in the seventeenth century in the context of the Copernican controversy. Carmelite provincial and supporter of Galileo, Paolo Foscarini, put it this way:

> Since something new is always being added to the human sciences, and since many things are seen with the passage of time to be false which previously were thought to be true, it could happen that, when the falsity of a philosophical opinion [to which the authority of scripture has been

37. Locke, *Essay concerning Human Understanding* 4.12.10, ed. Peter H. Nidditch (Oxford: Clarendon, 1979), 645. The contrast between knowledge and opinion goes back to Plato, *Republic* 478.

38. Locke, *Essay concerning Human Understanding* 4.12.11 (Nidditch edition, p. 646).

attached] has been detected, the authority of the Scriptures would be destroyed.[39]

Foscarini here questions the wisdom of attaching scriptural interpretation to past Ptolemaic and Aristotelian understandings of the cosmos. But his position about changing scientific claims is quite close to what contemporary philosophers of science refer to as "pessimistic meta-induction." This is the idea that what we learn from the history of science is that no scientific theory ever prevails in the long term. From historical examples we make an inductive generalization that leads us to be pessimistic about the truth-claims of current science. It follows that it would be reasonable to expect present scientific theories to be replaced in time.

Being on the Right Side of History

Before moving on to a discussion of how these principles might apply in the case of Christianity and evolutionary theory, it is worth briefly considering a few historical instances of religious resistance to scientific orthodoxy that might, with the benefit of hindsight, count as "justifiable conflicts." These examples highlight the shifting ground of scientific (or philosophical) orthodoxy, and that contesting a contemporary scientific consensus might subsequently turn out to be "correct" by later standards. Part of the complexity of the historical relations between science and religion lies in the fact that what counted as "science" in the distant past does not map directly onto our present conceptions of science. So it is important to note that before the scientific revolution of the seventeenth century what we would call "scientific orthodoxy" usually consisted in adherence to the natural philosophy of Aristotle.

A conspicuous and longstanding tension between "science" and Christianity concerns the issue of whether the universe has always existed, as philosopher Aristotle had taught. From the very inception of Christianity the fathers of the church rejected this claim, which was a common philosophical position, underscored by the authority of Aristotle. Indeed, resistance to this idea dates back at least to the Jewish philosopher Philo of Alexandria (25 BCE–50 CE), and continued through the Middle Ages. This rejection was

---

39. Paolo Foscarini, "Letter on the Motion of the Earth," quoted in Richard Blackwell, *Galileo, Bellarmine and the Bible* (Notre Dame: University of Notre Dame Press, 1991).

based on a commitment to the doctrine of creation and to the biblical story of human origins.[40] Since the middle of the last century the scientific consensus has shifted to the view that the universe began with the "big bang," and thus had a beginning in time or, perhaps more correctly, "with time." (This latter way of thinking about the creation of time itself is analogous to Augustine's suggestion that time was part of God's creation, rather than constituting a temporal frame within which creation took place.) So it turns out that the idea of a temporal universe is more consistent with contemporary "big bang" cosmology than was Aristotle's view of the eternity of the world. The point here is not to suggest some kind of "anticipation" of the modern position, and certainly not that the church fathers had some dim presentiment of big bang cosmology. Rather it is to suggest that a rigorous insistence on a perfect mapping of theological doctrines onto scientific orthodoxies will sometimes result in perverse outcomes.

A related example concerns the Christian philosopher John Philoponus (490–570) who opposed a number of the teachings of Neoplatonic commentators on Aristotle, including the doctrine of the eternity of the world. More importantly, perhaps, he also contested the common idea of the divinity of the heavens and the belief that the heavenly bodies might be moved by desire or by angelic agents. This was related to his rejection of Aristotelian theories of terrestrial motion, which required that objects in motion be constantly acted upon by a motive force. This refusal to accept the preeminent scientific authority of the time enabled Philoponus to ask whether God might have imparted a kinetic force to heavenly bodies so that they moved in a way that was akin to the motions of heavy and light bodies. In short, he proposed a unified theory of dynamics based on impetus. These ideas, preserved during the Middle Ages in Arabic works, influenced the new conceptions of motion that developed during the scientific revolution and eventually replaced those of Aristotle.[41] Some of Philoponus's motivation was theological; some was based on empirical evidence.

A further example of creative conflict came with the medieval condemnation of certain philosophical and scientific teachings, mostly those of Aristotle. On March 7, 1277, Stephen Tempier, the Bishop of Paris, con-

40. Thomas Aquinas allowed that God could have created an eternal universe, but he concluded that God had not. *De aeternitate mundi* 24. The point is that any created universe will depend on God for its existence, whether it is eternal or created in/with time.

41. On Philoponus, see Richard Sorabji, ed., *Philoponus and the Rejection of Aristotelian Science* (London: Duckworth, 1987); David C. Lindberg, *The Beginnings of Western Science* (Chicago: University of Chicago Press, 2007), 307–13.

demned 219 articles of theology and natural philosophy.[42] While there was considerable diversity among these 219 propositions, an assumption that lay behind at least some of the condemnations was the principle that God, on account of his omnipotence, could not be bound by the limits prescribed by Aristotelian doctrines. It followed that some states of affairs claimed to be "scientifically" impossible were in fact possible for God. It has been argued by some historians of science that these condemnations had the effect of liberating medieval thinkers from the prescriptions of Aristotle, allowing them to think counterfactually and hypothetically. Pierre Duhem went so far as to argue that this event signaled the beginning of the path to modern science. Few historians would now argue this without significant qualifications, but the event at least suggests that resistance to scientific orthodoxy (albeit defined in this instance as conformity with Aristotle) can sometimes have unintended positive outcomes.[43]

One last example is to do with the scientific revolution itself, which came about through a sustained and wholesale rejection of a range of scientific orthodoxies. There were multiple causal factors for this complex set of events but, again, religiously motivated opposition to some features of the prevailing consensus about scientific knowledge was important. I have written at length elsewhere about how the biblical doctrine of the Fall played a major role in the foundation and promotion of a new experimental approach to the natural world, but it is worth briefly retelling here.[44] The basic story is this. Aristotelian science had been based on commonsense observations of nature in its normal or customary state. The assumption was that human

42. For the text of the condemnation see David Piché, ed., *La condemnation parisienne de 1277. Texte latin, traduction, introduction et commentaire* (Paris: Vrin, 1999). See also Jan A. Aertsen, Kent Emery Jr., and Andreas Speer, eds., *Nach der Verurteilung von 1277. Philosophie und Theologie an der Universität von Paris im letzten Viertel des 13. Jahrhunderts. Studien und Texte* (Berlin: De Gruyter, 2001); John F. Wippel, "The Condemnations of 1270 and 1277 at Paris," *The Journal of Medieval and Renaissance Studies* 7 (1977): 169–201.

43. John E. Murdoch, "Pierre Duhem and the History of Late Medieval Science and Philosophy in the Latin West," in *Gli studi di filosofia medievale fra otto e novecento*, ed. Alfonso Maier and Ruedi Imbach (Rome: Edizioni di Storia e Letteratura, 1991), 253–302. Edward Grant, "The Condemnation of 1277, God's Absolute Power, and Physical Thought in the Late Middle Ages," *Viator* 10 (1979): 211–44. On the scientific relevance of divine omnipotence, see also Amos Funkenstein, *Theology and the Scientific Imagination* (Princeton: Princeton University Press, 1986), 10–12.

44. Peter Harrison, *The Fall of Man and the Foundations of Science* (Cambridge: Cambridge University Press, 2007); "Original Sin and the Problem of Knowledge in Early Modern Europe," *Journal of the History of Ideas* 63 (2002): 239–59.

minds and senses were naturally oriented toward knowledge, and that the operations of the natural world were transparent to human investigators. Following the Protestant reformation, however, a number of thinkers focused on the question of how theological anthropology (broadly, theological conceptions of the person) might make a difference to the pursuit of scientific knowledge. Specifically, they asked how the fallen condition of human beings might affect the acquisition of scientific knowledge. A number of pioneers of the new experimental approaches argued that while in the original creation human minds and senses could indeed have intuited the operations of nature in perfect detail (as Aristotle had thought), after the Fall this was no longer possible. The world itself, moreover, was also in a fallen state, rendering it opaque to human investigation. Commonsense observation of nature in its normal course was no longer enough. Instead, carefully designed and painstakingly conducted experiments were necessary to wring secrets from nature. These were to involve repeated trials and many observers; very often the results would be counterintuitive rather than commonsensical. Knowledge, moreover, was to be corporate and cumulative. Instruments such as the microscope and telescope were also necessary to augment senses that had been weakened by the Fall. This new experimental approach was no longer thought to give rise to the certain knowledge that for Aristotle had been the measure of true science. Rather, as Locke pointed out, experimental knowledge provided material convenience, but not certainty.

In sum, in the past religious tension with the prevailing scientific consensus has sometimes been productive of new scientific ideas and methods. Of course, it is not possible to specify in advance *which* instances of conflict are likely to be fruitful. Neither is it clear that the positive outcomes that resulted from these instances were intended by the relevant historical agents. But there is enough evidence to suggest that inflexible advocacy of an unquestioning harmony can be less than optimal for both science and religion. The difficulty, of course, is to discern when resistance to harmony is called for.

Christian Faith and Evolution

What follows from all this for the topic at hand? Here are two tentative prescriptions. First, perhaps the key principle that we learn from Augustine is that it is worth seriously considering whether resolving theoretical questions about the natural world is more important than seeking to understand the moral and religious purpose of our own lives. Augustine famously declared in the *Soliloquies* that he desired to know only God and the soul, and nothing more.[45] For Augustine, while knowledge of the natural world is a good thing, preoccupation with it is a clear example of a misplaced love—the kind of preference for lower goods that characterizes the fallen condition of human beings. Were Augustine to pass judgment on the priorities of our own age, he might suggest that our present veneration of scientific knowledge is a manifestation of the misplaced priorities that he constantly warns against. Augustine spoke, in terms now unfamiliar to us, about the vice of curiosity. But the warning here is against the fetishizing of certain forms of highly regarded intellectual activities. This, perhaps, is something that we can learn from religiously motivated opponents of evolution, however mistaken we may believe they are in their views about creation and how it relates to science.

This basic priority also puts into perspective Augustine's oft-repeated complaint about fellow Christians making fallacious scientific claims and claiming religious authority for them. This looks like a ready-made criticism of contemporary scientific creationists. It is that, and it is most often cited in that context. But placing this criticism within the broader scheme of Augustine's priorities, such activities, however lamentable and damaging to the credibility of Christianity, at least prioritize the things that are ultimately the most important. This is not to advocate an "anti-science" agenda, but rather to point to the significance of Augustine's diagnosis of the disordering of human "loves," and of the need to observe the correct priorities.[46]

The second lesson to be drawn from Augustine's treatment of these issues concerns the status of scientific knowledge and the principle of prudence. Augustine subscribed to a prevailing understanding of *scientia* according to which scientific knowledge must have the force of logical demonstration. However, from the eighteenth century onward no one has really thought that scientific knowledge could attain that level of certainty. Or perhaps I should say that no one with insight into how science really operates has thought that. As noted earlier, John Locke presciently observed that what we get from inductive knowledge is not truth but utility. But in any case, Augustine had also recognized that much current speculation about the natural world fell well short of the strict requirements of *scientia*. For this reason he counseled against too close an alignment of uncertain science with scripture, since subsequent reasonings or discoveries might put both in doubt. The

45. Augustine, *Soliloquies* 1.2.7.
46. Augustine, *On Christian Doctrine* 1.27.28; *Confessions* 4.10.15.

general point that need not be further labored is this: science changes. And it changes in ways that suggest it cannot be invariably truth-tracking.

Does this apply to evolutionary theory, and would it be rational to expect that at some future time it might be discredited? Here it is important not to be misled by generalizations about "science" and "scientific knowledge," particularly those that rely on physics as the basic model. Just as Aristotle and Augustine had rather different understandings of what counts as science from ours, even today a great variety of quite disparate disciplines crowd together under the umbrella of science, and their claims are justified in a variety of ways. The broad general claim of evolutionary theory—descent with modification—is really a historical claim about past events (although arguably descent with modification can be observed in the present). Such claims have a very different status from claims made, for example, in the sphere of quantum theory (which are puzzling but extremely well attested experimentally) or M-theory (which are not attested experimentally, but are consistent mathematically).[47] Historical claims fall well short of logical certainty, but for all that, many of them are so highly probable that it is not reasonable to doubt them—that there was a Great War in Europe between the years 1914 and 1918, or that there was a Roman Emperor, Julius Caesar, who lived in the first century BCE. Christianity also rests upon historical claims. Is it possible that at some future time it will be shown that there was no World War I, or Julius Caesar? This would seem extremely unlikely, but it depends on the strength of the evidence for the relevant events. So the principle of pessimistic meta-induction has less force in the case of the historical sciences than the theoretical sciences. Specifically, historical claims about an old earth and species change are far less susceptible to future falsification than more theoretical claims in the sphere of physics.

When we consider the specific *mechanisms* of evolution, however, to a degree we move out of the territory of historical claims. Here we see the kinds of competing and changing theoretical conceptions that characterize the history of other sciences. These concern discussions about the relative importance of natural selection, gene flow, genetic drift, developmental mechanisms, plasticity, niche construction, and the Lamarckian-sounding extra-genetic inheritance.[48] Even looking at the relatively short history of

47. Augustine himself actually discussed the problems of historical testimony in relation to demonstrated truths in *De fide rerum invisibilium* 2.4 (Patrologia Latina 40:173–74).

48. Kevin Laland et al., "Does Evolutionary Theory Need a Rethink?" *Nature* 514 (2014): 161–64.

evolutionary thinking, we have sufficient grounds to think that there will be future developments in these areas. The lesson here is that the prudential principle of Augustine, to do with premature commitment to particular scientific doctrines, should be applied in proportion to the degree of the probability of the various doctrines. This, of course, is often what is at issue. But consideration of the last 150 years suggests that the basic principle of descent with modification has stood the test of time (thus far), while claims about the mechanisms that underlie descent have changed. In particular, evolutionary processes may not be as random in their operations or outcomes as has often been supposed. This is not to concede much to young earth creationists, but is relevant to concerns about the apparent randomness and directionlessness of evolutionary processes.

Parenthetically, it is also important to distinguish the *mechanisms* of evolutionary theory from its putative moral and philosophical *implications*, which are far more contestable than the basic science. Even before the publication of Darwin's *Origin of Species* in 1859, evolutionary thinking had been freighted with philosophical and moral meanings that went well beyond its scientific import. In the nineteenth century the "epic of evolution" genre sought to imbue the evolutionary story with a mythic status that set it on the path of conflict with longstanding Judeo-Christian creation myths.[49] More recently, "Big History" and evolutionary psychology have sought to elevate evolution into an all-encompassing philosophy that can provide answers to any of life's questions that are considered to be worth asking.[50] These claims are extra-scientific accretions to the theory, but they are often conflated with the underlying science and imbued with its authority. What many fundamentalist critics of evolution are responding to is as much the moral package that they (mistakenly) believe is intrinsic to evolutionary theory as the theory itself. In relation to the question of conflict, then, it is important to be able to distinguish highly contestable implications of a theory from the more secure science that lies beneath those implications.

To return to the general point, if we consider again Augustine's principle (2a), the priority of demonstration, there will turn out to be no instances in which scripture need yield to demonstrative science in the way Augustine envisaged, because no science is demonstrative. That said, some historical

---

49. See, e.g., Ian Hesketh, "The Recurrence of the Evolutionary Epic," *Journal of the Philosophy of History* 9 (2015): 196–219.

50. Denis Alexander and Ronald Numbers, eds., *Biology and Ideology: From Descartes to Dawkins* (Chicago: University of Chicago Press, 2010); Ian Hesketh, "The Story of Big History," *History of the Present* 4 (2014): 171–202.

claims reach such a degree of probability that it would be irrational to deny them. The question is which of the claims made on behalf of evolutionary theory achieve that degree of certainty.

Conclusion

To conclude, in all of this I have been mostly considering one variable—science. Of course, religious doctrines also undergo change and development. And any discussion of the implications of evolutionary theory for traditional doctrines concerning the origin of sin and of human beings must closely scrutinize the history and status of those doctrines too. To some degree these will be susceptible to the kind of historical relativization that might be applied to scientific theories. But that is a discussion for another time.

It should be obvious that there are good reasons for thinking that hard irenicism has significant problems. On the foregoing analysis, this is largely because science cannot be directly equated with "truths about the natural world." From the fallibility (or, less pejoratively, mutability) of the natural sciences it follows that there is at least the possibility, on occasion, that some scientific claims will just happen to conflict with core Christian beliefs. It also follows that the presumption should not always be that it is Christian doctrines that need to be reformulated in light of new scientific claims. This presumption has the attendant danger that any state of affairs will be consistent with the tenets of Christianity, which would suggest (and here the logical positivists were on the right track) that the truth or otherwise of Christianity makes no conceivable difference to the way the world is. My other worry is that hard irenicism might be more a function of the present high epistemic status of scientific claims than of careful scrutiny of these claims and their implications. The fact is that science sometimes gets things wrong—in the long historical view, gets most things wrong—and a revised principle of prudence would suggest keeping a respectable distance between science and Christianity. In the specific case of evolution it cannot be doubted that the basic idea of descent with modification is well founded. But it is not a demonstrative truth. Neither does it follow that every aspect of the theory is well founded. And it is vital to distinguish well-founded theories from their broader implications, which may be more speculative and ideologically loaded.

But beyond these prudential considerations is Augustine's valuing of activities that conduce to the cure of souls and the love of God. This represents an overriding consideration that takes us beyond quarrels about facts concerning the physical world to an advocacy of dwelling upon what is excellent and praiseworthy. There is a danger that well-motivated attempts to establish peaceful relations between science and religion might neglect the more fundamental priority of which Augustine constantly reminds us. This is one of the incipient dangers of insisting that science and religion must always be unproblematically compatible.

Index of Names

Index of Subjects